Ethical Challenges in Global Public Health

Ethical Challenges in Global Public Health: Climate Change, Pollution, and the Health of the Poor

Edited by
Philip J. Landrigan and Andrea Vicini, SJ

⁕PICKWICK *Publications* · Eugene, Oregon

ETHICAL CHALLENGES IN GLOBAL PUBLIC HEALTH
Climate Change, Pollution, and the Health of the Poor

Copyright © 2021 Philip J. Landrigan and Andrea Vicini, SJ. All rights reserved. Except for brief quotations in critical publications or reviews, no part of this book may be reproduced in any manner without prior written permission from the publisher. Write: Permissions, Wipf and Stock Publishers, 199 W. 8th Ave., Suite 3, Eugene, OR 97401.

Pickwick Publications
An Imprint of Wipf and Stock Publishers
199 W. 8th Ave., Suite 3
Eugene, OR 97401

www.wipfandstock.com

PAPERBACK ISBN: 978-1-7252-9174-4
HARDCOVER ISBN: 978-1-7252-9173-7
EBOOK ISBN: 978-1-7252-9175-1

In gratitude to the Boston College community and to everyone who is promoting global public health.

Global Theological Ethics – Book Series

Series Editors
Jason King, St. Vincent College
M. Therese Lysaught, Loyola University Chicago

The *Global Theological Ethics* book series focuses on works that feature authors from around the world, draw on resources from the traditions of Catholic theological ethics, and attend to concrete issues facing the world today. It advances the *Journal of Moral Theology*'s mission of fostering scholarship deeply rooted in traditions of inquiry about the moral life, engaged with contemporary issues, and exploring the interface of Catholic moral theology philosophy, economics, political philosophy, psychology, and more.

This series is sponsored in conjunction with the Catholic Theological Ethics and the World Church. The CTEWC recognizes the need to dialogue from and beyond local cultures and to interconnect within a world church. Its global network of scholars, practitioners, and activists fosters cross-cultural, interdisciplinary conversations—via conferences, symposia, and colloquia, both in-person and virtually—about critical issues in theological ethics, shaped by shared visions of hope.

Online versions of the volumes in the *Global Theological Ethics* series are available for free download as chapters at jmt.scholasticahq.com. Paper copies may be purchased from Wipf & Stock. This dual approach reflects the *Journal of Moral Theology's* commitment to the common good as it seeks to make the scholarship of Catholic theological ethicists broadly available, especially across borders.

Series Titles

Ethical Challenges in Global Public Health: Climate Change, Pollution, and the Health of the Poor, edited by Philip J. Landrigan and Andrea Vicini, SJ (2021)

Contents

Introduction
Global Public Health and the Promotion of the Common Good
Andrea Vicini, SJ 1

Part 1: Setting the Context
Health and Climate Change
Walter Ricciardi and Laura Mancini 14

The Vital Contribution of Independent, Ethically Grounded Research to the Global Health Agenda
Kurt Straif 28

Part 2: The Changing Context of Global Public Health
Challenges Confronting Global Public Health
Keith Martin. 40

Pollution, Climate Change, and Global Public Health: Social Justice and the Common Good
Philip J. Landrigan. 53

Global Public Health and Catholic Insights: Collaboration on Enduring Challenges
Michael D. Rozier, SJ. 63

The Affordable Care Act and Pharmaceuticals: An Economic Perspective
Tracy L. Regan. 75

Part 3: Global Public Health Ethics
Social Structures and Global Public Health Ethics
Daniel J. Daly. 84

Ethics and Equity in Global Health: The Preferential Option for the Poor
Alexandre A. Martins 96

Social Justice and the Common Good: Improving the Catholic Social Teaching Framework
Lisa Sowle Cahill 106

Part 4: International Approaches to Global Public Health
Challenges in Global Health, Culture, and Ethics in Africa
Jacquineau Azetsop, SJ. 119

Public Health Concerns in India: A Review
Stanislaus Alla, SJ 131

A European Take on Global Public Health: Applying the Catholic Principle of Subsidiarity to Global Health Governance
Thana Cristina de Campos. 141

Part 5: Building an Ethical Framework for Education and Research in Global Public Health
Inequities as an Ethical Imperative: Challenges Related to Identification, Engagement, and Interventions in Minority Health
Nadia N. Abuelezam. 153

Addressing Health Disparities Among Families: Policy Approaches to Improve Infant Health
Summer Sherburne Hawkins. 165

Humanitarian Aid, Infectious Diseases, and Global Public Health
Nils Hennig 181

Conclusion
An Ethical Agenda for Global Public Health
Paul E. Farmer and Andrea Vicini, SJ 193

Introduction

Global Public Health and the Promotion of the Common Good

Andrea Vicini, SJ

WITH THE CATCHY TITLE of "Can public health save the world?" for its July 2020 COVID-19 special issue, the *Hopkins Bloomberg Public Health* magazine explored various issues raised by the global pandemic caused by the coronavirus called COVID-19.[1] To answer, the magazine's articles spanned from the past to the present and then speculated about the future. First, authors focused on what led to the global pandemic—the prequel—with the historical lessons unlearned and the public health policies underfunded that facilitated the spreading of the pandemic.[2] Second, concentrating on the present, on what the magazine described as "the fight," the authors considered what was occurring and how effective it might be.[3] Third, while thinking about the future might

---

[1] See Ellen J. MacKenzie, "Let's Fix Things for Good: COVID-19 Is Teaching Us a Brutal Lesson: Invest in Public Health or Suffer the Consequences," *Hopkins Bloomberg Public Health*, COVID-19 Special Issue (2020): 1–2.

[2] See Tom Inglesby, "Never Rest: Big Biological Threats over the Last Couple of Decades Have Taught Us One Thing: More Are on the Way," *Hopkins Bloomberg Public Health*, COVID-19 Special Issue (2020): 7; Carrie Arnold, "Caught Off Guard: How Policies for Preparedness Could—and Should—Have Protected Us," *Hopkins Bloomberg Public Health*, COVID-19 Special Issue (2020): 8–9.

[3] See Carrie Arnold, "Countering the Infodemic: Misinformation About SARS-CoV-2 Is as Contagious as the Virus Itself," *Hopkins Bloomberg Public Health*, COVID-19 Special Issue (2020): 24–25; Carrie Arnold, "The Natural Fix: An Old-School Approach—Using Antibodies from COVID-19 Survivors—May Be a Fast, Stop-Gap Solution for a Modern Pandemic," *Hopkins Bloomberg Public Health*, COVID-19 Special Issue (2020): 31–33; Jackie Powder, "Coping with COVID-19: A Global Approach to Universal Psychological Responses," *Hopkins Bloomberg Public Health*, COVID-19 Special Issue (2020): 37; Christen Brownlee, "A Crisis within a Crisis: The Pandemic Has Created a Convergence of Suicide Risk Factors That Also Need a Public Health Response," *Hopkins Bloomberg Public Health*, COVID-19 Special Issue (2020): 38–39; Cathy Shufro, "Breaking the Chain: One COVID-19 Patient Could Lead to Thousand New Cases: Contact Tracers Use Calls, Texts, and Personal Persuasion to Prevent That from Happening," *Hopkins Bloomberg Public Health*, COVID-19 Special Issue (2020): 13–15; Karen Kruse Thomas and Dayna Kerecman Myers, "Racism and COVID-19," *Hopkins Bloomberg Public Health*, COVID-19 Special Issue (2020): 16–17; Laura Wexler and Brennen Jensen, "Voices of the Vulnerable: For Asylum Seekers, the Incarcerated, Frontline Doctors, and Others,

generate anxiety because of unforeseen factors and the uncertainty that they generate, the authors felt it was necessary to investigate where planning might still be insufficient.[4]

With no ambition of covering all the needed topics, and by articulating concerns and methodological approaches raised by the COVID-19 global pandemic, the magazine reminded us that any issue should be examined and addressed by focusing on those affected, on who has the competence and responsibility to intervene, and on who concretely is engaged to promote health on the ground. The collection of articles called us to learn from the past, critically question the present, and look to the future we envision. Furthermore, it showed us to be grateful for all those who dedicate themselves to care for the sick and to promote health, especially because many healthcare professionals suffer mental, emotional, and physical burnout while dealing with the protracted health emergencies that characterize any global pandemic.

The global pandemic has affected millions of people, taken innumerable lives, unveiled the limits and vulnerabilities of health systems both in the Global North and in the Global South, challenged the world economy, harmed educational enterprises, and tested the human ability to adapt to changed living conditions that limit and inhibit social interactions. While global public health has played an important role in assuring living conditions on the planet, skeptics, whether because of culpable ignorance or misguided political biases, may still disregard a balanced assessment of how global public health enriches, strengthens, and expands the promotion of health by focusing on the health of populations and of the whole world.

While we avoid referring to global public health as our savior, we acknowledge and appreciate its positive contributions in promoting sustainability and reinforcing human resilience on earth. We accept that global public health cannot save the world alone but joins many other disciplines engaged in research and social transformation—from multiple natural sciences to social sciences, from the humanities to

---

COVID-19 Has Made Hard Lives Harder," *Hopkins Bloomberg Public Health*, COVID-19 Special Issue (2020): 18–21.

[4] See Karen Blum, "Fast Science: COVID-19 Research Is Happening at Lightning Speed—Sometimes at the Expense of Sound Science," *Hopkins Bloomberg Public Health*, COVID-19 Special Issue (2020): 44–45; Ronald J. Daniels, "A Vital Mission: Universities Responded to the Pandemic with Sound Science and Advice: We Can Still Do More," *Hopkins Bloomberg Public Health*, COVID-19 Special Issue (2020): 43; Julie Scharper, "The Vaccine Challenge: A Return to Normal Requires a Vaccine for SARS-CoV-2: What Will It Take to Create One and Get It to Those Who Need It Most?" *Hopkins Bloomberg Public Health*, COVID-19 Special Issue (2020): 46–47; Jackie Powder, "Envisioning a Post-Pandemic World: How COVID-19 Has Reset the Present and the Future," *Hopkins Bloomberg Public Health*, COVID-19 Special Issue (2020): 48–49.

political sciences and religion. Isolated and disconnected efforts are praiseworthy but insufficient. What is needed are collaborative efforts that promote multidisciplinary participation and aim at offering realistic and appropriate solutions to complex problems.

The COVID-19 global pandemic invites civil society to prioritize global public health by continuing to invest in research and offering healthcare services to every citizen, particularly those who are more vulnerable; to secure jobs while extending unemployment benefits and providing economic support to individuals, families, and struggling economic activities; and by spending what is needed to safely reopen educational institutions.

GLOBAL PUBLIC HEALTH AS A COMMON GOOD

At the core of any reflection regarding global public health is the profound conviction, tested by everyone's experience, that health is a good, both for human beings and the whole planet. Moreover, health is a shared, common good that concerns everyone. However, health is a fragile and vulnerable good that demands care by examining who lacks health, by asking why people are not healthy, by investigating what affects their well-being and flourishing, and by implementing what aims at restoring health as much as possible for populations and the whole humankind.

These are very simple statements. Hopefully, they are straightforward and shared across cultures, political and religious beliefs, as well as ethnic and linguistic differences. Such a global understanding of health as a common good for humankind and the globe requires a comprehensive approach. As the science and art of promoting good health, preventing disease, and extending longevity in countries around the world in very inclusive manners, global public health fulfils such an ambitious scope by aiming at equity and justice in health and within society.

By examining the social determinants of health, and the role that they play in shaping health outcomes, global public health is rooted in and aims at promoting social justice. A substantial body of scholarship in global public health elucidates the social, political, economic, and environmental factors that influence patterns of health and disease and that drive disparities and inequities in health. Despite the clear connections between global public health and social justice, there has been surprisingly little scholarly exploration of the ethical challenges confronting global public health. This volume engages this gap.

AN INCLUSIVE COMMON GOOD

By aiming at promoting health as a common good, global public health is not a pseudo-global meta-narrative that imposes a vision of health chosen by dominant powers—whether cultural, economic, political, or religious. On the contrary, any true and realistic understanding of health as a common good is necessarily inclusive and, even more, focuses on those who are left on the margins, who are not considered relevant interlocutors and participants.

In theological ethics, David Hollenbach, SJ, defines the common good "as an ensemble of goods that embody the good of communion, love, and solidarity to a real though limited degree in the multiple forms of human interaction."[5] Health fits well among these goods because it promotes both individual and social flourishing. In today's pluralistic societies, the "pursuit of the common good demands full respect for the many different forms of interrelationship and community in which human beings achieve the good in history."[6] Hence, health as a common good does not imply a vision of health that aims at a disembodied perfection influenced by ideological misconceptions, as in the case of the twentieth-century eugenics or the more recent liberal eugenics. In other words, health does not exclude the limitations and disabilities that accompany human existence and includes any type of diversity—ethnic, racial, cultural, political, and religious.

Both in the Global North[7] and in the developing Global South,[8] unjust inequities characterize and plague the social, economic, and political contexts, as well as health systems and their services.[9] Thus, the common good is closely connected to social justice and equality. By stressing the preferential option for the poor,[10] the common good aims at greater equality by requiring a resolute and effective commitment to reduce and, hopefully, eliminate the causes of unjust inequalities and to promote health at a global level.

In the tradition of Catholic reflection, the common good depends both on the Christian faith, which is concerned with the good of each one, and on the rational reflection on human experience, shared by

---

[5] David Hollenbach, *The Common Good and Christian Ethics* (Cambridge, UK: Cambridge University Press, 2002), 136. Quoted in Gonzalo Villagrán Medina, SJ, "Iglesia y Vida Pública en David Hollenbach: Aproximación a Su Método en Teología Moral," *Theologica Xaveriana* 64, no. 177 (2014): 241–266, at 247.

[6] Hollenbach, *The Common Good*, 136.

[7] See Kate Ward and Kenneth R. Himes, "'Growing Apart': The Rise of Inequality," *Theological Studies* 75, no. 1 (2014): 118–132; Kate Ward and Kenneth R. Himes, ed., *Growing Apart: Religious Reflection on the Rise of Economic Inequality* (Basel, Switzerland: MDPI, 2019).

[8] See Agbonkhianmeghe E. Orobator, "*Caritas in Veritate* and Africa's Burden of (under)Development," *Theological Studies* 71, no. 2 (2010): 320–334. In this book, see the chapters of Jacquineau Azetsop and Stanislaus Alla.

[9] See National Academies of Sciences Engineering Medicine, *Crossing the Global Quality Chasm: Improving Health Care Worldwide* (Washington, DC: The National Academies Press, 2018).

[10] In this volume, the chapters by Michael Rozier, Alexandre Martins, Lisa Sowle Cahill, as well as Paul Farmer and Andrea Vicini further discuss the preferential option for the poor.

each person regardless of any cultural, religious, linguistic, racial, social, and political difference. Hence, the common good is, at the same time, specific to the Catholic and Christian tradition and integral to basic human experience.

Concretely, the common good presupposes the right to health for every citizen—regardless of income, social location, capabilities, or working skills—and calls each person to contribute to the realization of the common good by promoting health. Further, health depends on personal, local, national, and global involvement, from those who are directly engaged in promoting health (i.e., doctors, nurses,[11] technicians, and administrators) to politicians, legislators, and governments (responsible for the development of the health system in each country) to groups, organizations, foundations and institutions that are at the service of health at a global level (e.g., Partners in Health, Médecins Sans Frontières, Bill and Melinda Gates Foundation, the Centers for Disease Control and Prevention, and the World Health Organization) and, finally, each citizen. Among these social actors, to stress its commitment to promote the common good in the health sector, the December 2016 issue of *Health Progress*, published by the Catholic Health Association—the "largest group of nonprofit health care providers in the nation" serving "more than 600 hospitals and 1,600 long-term care and other health facilities in all 50 states"[12]—dedicated the entire issue to the common good.[13]

RATIONALE AND OUTLINE

Initially a conference organized and held at Boston College (Boston, Massachusetts) in September 2019, this volume gathers almost all the original contributions, in a revised and expanded form. By examining the collection of essays, global public health emerges as a complex discipline that requires multidisciplinary contributions—from ethics to economics and public policy, from nursing to social work, from medicine to population health—to address the social determinants of health and to articulate transformative practices and structures able to improve the quality of life and foster health for individuals, communities, and the whole planet.

The choice of topics discussed does not have the ambition of being complete and exhaustive. Hopefully, both expert readers and

---

[11] See Elma Lourdes Campos Pavone Zoboli, "Cooperar Para el Bien Común: ¿Responsabilidad Social de la Enfermería?" *Bioethikos* 1, no. 1 (2007): 118–123.
[12] Catholic Health Association of the United States, "About" (2020), www.chausa.org/about/about.
[13] As examples, see Meghan Clark, "Health Equity, Solidarity and the Common Good: Who Lives, Who Dies, Who Tells Your Story," *Health Progress* 97, no. 6 (2016): 9–12; Thomas Nairn, "Health Care Decisions for the Common Good," *Health Progress* 97, no. 6 (2016): 4–7; Deborah M. Spitalnik, "Disability Rights and the Common Good," *Health Progress* 97, no. 6 (2016): 48–53.

practitioners will appreciate the insights and experience shared, the rigor in engaging issues, and the concern for promoting ethical commitments that will benefit the earth and humankind, particularly those who are more vulnerable. At the same time, those familiar with the field of global public health and its ongoing ethical challenges will agree that global public health is strengthened by a robust ethical agenda.

The contributors were chosen, first, to represent and exemplify the diverse and multidisciplinary engagement in global public health at Boston College while, second, interacting with selected and outstanding scholars and activists, both nationally and internationally. Striving to be global is demanding. Being rooted in the scholarship and research that occurs in a particular context like Boston College, while attracting contributions from diverse locations in the US and abroad across continents, is one way to begin addressing multiple ethical issues in global public health.

In part one, "Setting the Context," the book explores two key ethical challenges in global public health by focusing, first, on the inseparable connection between the environment and health in times of climate change. Walter Ricciardi and Laura Mancini examine the problematic consequences on human and planetary health caused by the ongoing changes in our global climate: from direct to indirect effects on people, animals, and ecosystems, as well as from increased diseases to implications for mental health. Interventions to address the climate change crisis, and protect the quality of life on the planet, are needed and urgent. Restorative actions are required, and resolute political commitments should promote them. The international agreements that the authors discuss are one example that demands implementation.

Second, Kurt Straif further defines the context of the global health agenda by highlighting the important role played by accurate assessments of the negative impact on health—for individuals, communities, and the whole planet—caused by the production, use, and disposal of chemical products. His research and engagement with the Monographs Programme, promoted by the International Agency for Research on Cancer, allows us to appreciate the needed and vital contribution of independent, ethically grounded research that assesses the health risks of exposure to old and new chemicals, particularly in the case of cancers. At the same time, from an ethical standpoint, Straif shows how scientific research aimed at promoting global public health might expose the role and responsibility of multinational corporations in producing and distributing what could harm human and planetary health and might face corporate attempts to silence and stifle beneficial and groundbreaking investigative research.

In part two, four contributors reflect on "The Changing Context of Global Public Health" and its ethical implications. First, Keith Martin highlights the "Challenges Confronting Global Public Health" and presents solutions currently implemented and needed. Pandemics caused by infectious diseases dominate his analysis and, as a solution, he discusses the Global Health Security Agenda platform that allows the international community to collaborate, prevent, detect, and respond to disease outbreaks. At the same time, he reminds us of the gravity of non-communicable diseases (i.e., cardiovascular disease, cancers, respiratory diseases, diabetes, mental health, and injuries) that should be addressed by comprehensive public health programs able to foster prevention and promote adequate interventions and by focusing on the social determinants of health. Furthermore, he considers environmental threats and climate change, with the still insufficient commitment to address them. Martin concludes by discussing more issues that require significant commitments: from adequate financing for public health programs to political governance at the service of the citizens' needs, from avoiding corruption to fostering women's health and healthy nutrition.

Second, in his chapter "Pollution, Climate Change, and Global Public Health: Social Justice and the Common Good," Philip J. Landrigan summarizes current knowledge of the known and projected health effects of pollution and climate change to planetary health and examines the distribution of their impacts through the lens of social justice. In the world today, pollution is the largest environmental cause of disease, disability, and death—whether we consider pollution in air, oceans, and soil or caused by chemicals. At the same time, global climate change not only has numerous negative effects on the planet's ecosystems but also multiple adverse effects on human health. Furthermore, both pollution and climate change disproportionately affect the poor and the vulnerable and, among them, children and people living in the Global South and in poor communities worldwide. Hence, pollution, poverty, poor health, and lack of social justice are closely intertwined.

Third, Michael D. Rozier, SJ, focuses on "Global Public Health and Catholic Insights: Collaboration on Enduring Challenges" by showing how resources that inform the Roman Catholic ethical tradition could be valuable for global public health by helping to cultivate a sense of vocation among public health professionals, like the awareness of vocational commitments enjoyed in other healing professions. Moreover, the social teaching of the Catholic Church, particularly the preferential option for the poor, could help promote a more just distribution of global resources. Finally, dignity and solidarity—which inform and shape a Catholic approach—could provide the conceptual grounding needed to invest more energy in capacity building in low-resource settings and, at the same time, promote changes within the Church itself, empower communities in the Global South, and facilitate living lives nourished by joy and purpose. Hence, for Rozier, a more intentional relationship between global public health and the Catholic Church, with its highlighted teachings and commitments, would benefit both and particularly the people they aim to serve.

Fourth, in "The Affordable Care Act and Pharmaceuticals: An Economic Perspective," Tracy L. Regan discusses ongoing challenges and transformations in the American healthcare context after the 2010 Affordable Care Act by comparing it with what occurs in the United Kingdom, where the National Health Service is both the payer and provider of health care, and in France, where health coverage is universal and compulsory. In her assessment, the US health care system fails in basic dimensions—like preventative care and reimbursement schemes for physicians—but its innovation, technology, and research enabled many people to live longer and healthier lives. Moreover, in the US, regulation of the pharmaceutical industry and of drug prices should be part of the necessary reform of the health care system, but the political scene and market dynamics make any attempt to reform a currently impossible task.

In part three, three authors articulate their "Global Public Health Ethics." First, in his chapter on "Social Structures and Global Public Health Ethics," Daniel J. Daly relies on critical realist social theory because it provides an account of social reality that enables global public health ethicists to understand the causal mechanisms that perpetuate the suffering of the poor. Then, he ethically describes social structures by examining structures of virtue and vice. Finally, such an approach allows him to critically discuss two ongoing global public health crises—global warming and the lack of health workers in the Global South—by stressing how structures influence the moral character of individuals and produce social outcomes that promote or undercut human well-being. Notably, vicious structures foster social injustice and undermine the common good.

Second, in writing on "Ethics and Equity in Global Health: The Preferential Option for the Poor," Alexandre A. Martins denounces poverty as the main cause of health issues, diseases, and premature death. To break the vicious cycle caused by poverty, which begins with injustice and ends with death, he argues for an approach *from below*, from the experience of the poor, which places the voices and experiences of poor people at the center of discussions and actions in global public health by stressing the need of a preferential option for the poor. As an existential commitment, and an ethical imperative that inspires decision-making processes and informs concrete choices, such an option can greatly contribute to promote equity in global health. The samples of voices of poor people featured in his chapter exemplify his approach and engage us in searching for ways and practices that will break the unjust vicious cycle of poverty, vulnerability, lack of healthcare, and premature death.

Third, by writing on "Social Justice and the Common Good: Improving the Catholic Social Teaching Framework," Lisa Sowle Cahill recapitulates the modern history of Catholic social teaching by focusing on the common good and, particularly, the universal common good, as the indispensable criterion of social justice because it entails the equal participation of every member of society in basic material, social, and political goods. However, the promotion of the common good is challenged by the absence of real *political will*, the urgency of *conversion* of imaginations and worldviews, and the need to foster a *decentralized* global socioeconomic and political agency centered on popular mobilization. Additionally, the enduring secondary status of women in virtually every society urgently demands gender equality in having access to health resources and the social determinants of good health. Her chapter concludes with a case study from the Peruvian Amazon that exemplifies the commitment to promote the common good and the empowerment of women.

Part four offers three contributions that further enlarge the horizon of inquiry with "International Approaches to Global Public Health" from three continents: Africa, Asia, and Europe. First, Jacquineau Azetsop, SJ, guides us to explore the "Challenges in Global Health, Culture, and Ethics in Africa." An African perspective stresses how global health implies a vision of what it means to be a community that upholds human rights, social justice, respect for others, and health equity as necessary to live a good life. The realization of such a vision calls for a global solidarity that goes beyond national and continental boundaries. However, to implement this vision, most African countries face three major challenges: the fragmentation and pitiful shape of the healthcare system; the lack of real democracy with its consequences on public health policy and leadership; and the cultural inadequacy of the ethical principles regarding research and clinical work. In particular, health sector policy and planning, as well as the role of external partners in promoting health development and in implementing health system structures, are needed to overcome the fragmentation and inefficiency of health systems. Moreover, to shape an ethical approach informed by African contributions, he proposes *four contextualized principles:* the principle of respect for persons and for the alterity of their culture; the principle of social justice; the principle of public benefits; and the principle aimed at promoting local capacity building.

Second, Stanislaus Alla, SJ, reviews "Public Health Concerns in India." Diversity and plurality define India and such complexity is shared by several Asian nations. Cultural components, poverty and population concerns, illiteracy and ignorance, superstition and corruption require urgent attention, and they severely limit any effort to make healthcare accessible to large sections of the population. Moreover, as in other places in the world, in India climate change and pollution are devastating the lives of the poor. Alla begins his chapter by describing the state of people's health as a partial story of success. He stresses how healthcare has been made accessible to several sections of the Indian population; mother and child mortality rates have been reduced; and the average life expectancy has improved. Then, he critically examines what ails healthcare services in India by highlighting three concerns: the need for more funding devoted to public health programs; the challenge of eliminating food and water contamination; and the lack of political will in maintaining and expanding governmental health centers. Moreover, he proposes that public discourse on health should address the conflict between constitutional and cultural values. Finally, human rights should be considered not simply a legal concept but also a moral compass. This could help in promoting health by defining what is required to foster health and by prohibiting what is harmful for peoples' health.

Third, in her chapter "A European Take on Global Public Health: Applying the Catholic Principle of Subsidiarity to Global Health Governance," Thana Cristina de Campos contrasts two radically different approaches to governance in public health emergencies. On the one hand, she discusses a centralized approach that would further empower the World Health Organization (WHO). On the other hand, she presents a decentralized approach to global public health that is shaped by the principle of subsidiarity. Originally proposed in Catholic social teaching, such a principle informs governance within the European Union. The principle of subsidiarity establishes that when families, neighborhoods, and local communities can effectively address their own problems, they should do so; and only when they cannot, governments and other higher-level structures of power and authority should intervene and provide aid. For de Campos, fostering a decentralized subsidiarity is a promising principle for global health governance. Moreover, this principle justifies certain limitations of the WHO and of other higher-level global health authorities and powers by respecting the participation of local communities.

By studying three continental examples, these three chapters explore key elements in global public health and indicate how striving to promote global public health demands familiarity with the specificity of each local context. In other words, global public health appreciates and depends on particularity. Hence, the engagement of the diverse local contexts—with its citizens, cultures, religions, and institutions – is integral to fostering global public health.

In part five, three chapters contribute to "Building an Ethical Framework for Education and Research in Global Health." First, Nadia N. Abuelezam focuses on "Inequities as an Ethical Imperative: Challenges Related to Identification, Engagement, and Interventions in Minority Health." Health inequities are rooted in injustice, are often difficult to ameliorate, and require structural changes. In the American healthcare system, examples of inequities include maternal and infant health as well as lacking access to healthcare and health insurance. To address health inequities in the changing demographic of American society, she proposes a threefold approach that focuses, first, on the identification of minority populations to understand and document health inequities, unveil their risk factors and health outcomes, and develop research programs in minority health. Second, engagement with community members is required to understand the needs of minorities. Research should prioritize vulnerable populations and provide employment opportunities for them, like the Program in Community Engagement that worked with Black and Latinx men to ensure appropriate education around HIV prevention strategies. Finally, interventions are needed to address inequities in housing, employment, and education to improve health. Communities of opportunity exemplify this approach by focusing on children of low socioeconomic backgrounds in areas historically disadvantaged; by ensuring that healthcare professionals privilege prevention in providing care; and by empowering policymakers with information about the health inequities occurring in those communities.

Second, in her chapter "Addressing Health Disparities Among Families: Policy Approaches to Improve Infant Health," Summer Sherburne Hawkins shares the results of her evidence-based research showing that fiscal policies, particularly taxes, have downstream effects on infant health by influencing parental health behaviors as well as women's health habits during pregnancy. As a case study, state cigarette tax increases led to the largest benefits for the most vulnerable mothers and infants by proving that tobacco control policies, particularly cigarette taxes, reduce prenatal smoking and improve birth outcomes among the most vulnerable infants. Hence, an evidence-based approach to population health could promote needed policies aimed at improving the health and well-being of the most vulnerable children and families.

Third, in "Humanitarian Aid, Infectious Diseases, and Global Public Health," Nils Hennig examines ethical challenging priorities in humanitarian aid, global trends in both common and neglected infectious diseases, and key ethical issues in global health research. Considering his experience in critical contexts across the globe, he lists ten priorities that describe what humanitarian interventions should provide: (1) initial assessment of the situation, (2) water and sanitation, (3) food and nutrition, (4) shelter and site planning, (5) health care in the emergency phase, (6) control of communicable diseases and epidemics, (7) measles immunization, (8) public health surveillance, (9) human resources and training, and (10) coordination. The complexity of these needs reveals how humanitarian aid workers constantly face ethical challenges during the emergency phase while they aim at promoting health and preventing disease with equity.

Moreover, globally, and particularly in low-income countries, well-known infectious diseases are resurgent, spreading more rapidly than ever before, and new infectious diseases are being discovered at a higher rate than at any time in history. In terms of global health research, inequities regard who will set the research agenda, who will benefit from the research's results, and how will communities be involved or affected. Thus, in humanitarian interventions, the ethical decision-making framework should be informed by principles. Hennig proposes seven principles based on the World Health Organization's Global Health Ethics Unit recommendations: justice or fairness, beneficence, utility, respect for persons, liberty, reciprocity, and solidarity.

Finally, in the conclusion, Paul E. Farmer and Andrea Vicini, SJ, propose that "An Ethical Agenda for Global Public Health" be centered on the preferential option for the poor. It would animate global public health by caring for the well-being of everyone and for justice. It would unmask past and present attempts that undermine such an option. When the option for the poor is central, positive results are remarkable, for people in need and for the whole society. The constructive engagement of universities, with their research agendas and teaching commitment to education and formation, exemplifies how social transformation and the promotion of the common good can occur and help humanity and the planet to flourish. M

Andrea Vicini, SJ, is Michael P. Walsh Professor of Bioethics and Professor of Moral Theology in the Theology Department at Boston College. He is an alumnus of Boston College (STL and PhD) and holds a MD from the University of Bologna and an STD from the Pontifical Faculty of Theology of Southern Italy in Naples. At Boston College, he was Gasson Professor and taught in the School of Theology and Ministry. He also taught in Italy, Albania, Mexico, Chad, and France. He is co-chair of the international network Catholic Theological Ethics in the World Church, as well as lecturer and member of associations of moral theologians and bioethicists in Italy, Europe, and the US. His research interests and publications include theological bioethics, global public health, new biotechnologies, environmental issues, and fundamental theological ethics.

Part 1:
Setting the Context

Health and Climate Change

Walter Ricciardi and Laura Mancini

CLIMATE CHANGE THREATENS TO undermine the health gains of the last half century. One of the biggest challenges worldwide is the protection of public health, now severely threatened by globalization and urbanization. The anthropic impact is changing the environment on a global scale. These changes can have direct and indirect effects on the health of populations, also introducing new pathologies. We are now in the Anthropocene era where, for the first time, humans are changing the earth on which we live and affecting the creatures with which we share it. Climate change is already a large-scale global issue, increasingly affecting the health of all, particularly people at greatest risk of its adverse health effects, including children, the elderly, and other vulnerable population groups (i.e., immigrants, people with precarious housing conditions, and those chronically ill).[1]

Climate change threatens our health—whether we live in a rural village or on a small island, in coastal areas or in a large city; everyone is at risk. The scientific community is united in declaring that the effects of climate change on health, whether direct or indirect, are the most urgent public health problem to be faced today. The World Health Organization (WHO) argues that the health effects of climate change expected in the future, in particular those due to the progressive warming of the planet, will be among the most significant health problems to be addressed.

The next ten years will prove crucial for the health of both humans and the planet. Our emissions of carbon dioxide are driving up global temperatures, our exploitation of the environment is destroying biodiversity, our reckless use of chemicals threatens the insects on which we depend to pollinate crops, and our overconsumption of antibiotics is fueling the rise of antimicrobial resistance. According to WHO estimates, climate change will cause an additional 250,000 deaths worldwide each year between 2030 and 2050.[2] The impacts and consequences on human health are dramatic: as the WHO states, vector-related diseases will increase with higher humidity and

---

[1] See Philip J. Landrigan et al.,"Pollution and Children's Health," *Science of the Total Environment* 650, no. 2 (2019): 2389–2394.
[2] See World Health Organization, "Climate Change and Health," Modified February 1, 2018, www.who.int/news-room/fact-sheets/detail/climate-change-and-health.

temperatures, food production will be destabilized by drought, air pollution will lead to an increase in allergies and asthma, and warmer waters and floods will increase the risk of waterborne diseases. Climate changes affect social and environmental health determinants such as clean air, ecosystem health, safe drinking water and sufficient healthy food. Temperature-related death and illness, extreme events, and polluted or stressed ecosystems represent important issues of increasing concern with both health and economic consequences.

Protecting human health from climate change requires management at many levels, from the scientific assessment of risks and exposures for human populations to social, economic, and political aspects. It requires actions like application of the adaptation strategy at all levels of governance; strong support for research on adaptation to climate change to fill the gaps in our knowledge of the effects on ecosystems and health; coordination of adaptation policies; adoption of an ecosystem-based approach and the use of green and/or blue infrastructure in the development of adaptation actions; implementation of rapid alarm systems; and better harmonization and collaboration between the health and environmental sectors. The European Union's strategy on adaptation to climate change is an example of concrete actions to be implemented. The strategy aims to make Europe more resilient to climate change; in fact, by adopting a coherent approach and providing better coordination, it aims to improve the preparation and capacity of all levels of governance to respond to the effects of climate change like heat waves, floods, water scarcity, fires, coastal erosion, the appearance of invasive species, and other effects.

The health impact of climate change is based on four macro-classifications. First, climate change directly affects people and animals with diseases mainly related to homeostatic alteration of human and animal pathophysiology determined by change in the frequency of weather conditions and extreme events. Changes include the quality of livestock production. Non-communicable diseases increase with polluting agents. According to the WHO, seven million people die prematurely each year due to non-communicable diseases caused by pollution. Many atmospheric contaminants are also closely related to climate and climate change.[3] The WHO reports that only 12 percent of large cities are in compliance with the threshold values for air quality, and thus it recommends urgent action to reduce urban pollution.

Second, climate change indirectly impacts health effects because of the presence of vectors or vehicles of mainly infectious diseases. These include tick-borne encephalitis, Lyme borreliosis, malaria,

---

[3] See Raquel A. Silva et al., "Future Global Mortality from Changes in Air Pollution Attributable to Climate Change," *Nature Climate Change* 7 (2017): 647–651.

West Nile fever, and the whole group of infectious encephalitis, many respiratory diseases from hantavirus, chikungunya, dengue, leishmaniasis, and diseases related to the recently identified Zika virus. The effects also include epidemics caused by diseases previously limited to ecosystems that are peripheral with respect to areas of greater anthropization. Higher temperatures, milder winters, and wetter and warmer summers are expanding the areas where disease-carrying arthropods (e.g., ticks and mosquitoes) survive and multiply.

Third, climate change affects mental health. During an investigation (i.e., data from the US Behavioral Risk Factor Surveillance System, the largest database in the world on the subject), researchers at the Massachusetts Institute of Technology (MIT) observed that, with the general increase by one degree Celsius, medium-severity psychological pathologies were enhanced by 2 percent. The diseases detected included depression, states of anxiety, insomnia, fears, and generalized mental illness. Moreover, the increased probability of catastrophic events due to climate change (e.g., floods, fires, and progressive loss of arable land) could generate or exacerbate the reactions of already fragile individuals. Furthermore, the progressive reduction of animal and plant biodiversity, together with variation of the usual seasonal atmospheric parameters, leads to the perception of a state of imbalance that can induce or at least exacerbate even mild pathological conditions.

Fourth, climate change increases transmissions of diseases from animal to humans. Zoonoses, or infectious diseases transmitted by animals to humans (i.e., caused by bacteria, viruses, parasites, or prions), are a category of diseases strongly influenced by climate change. Although current knowledge does not allow general predictions of the impact of climate change on zoonoses, for some of these diseases the available evidence is varied and solid. This is the case of zoonoses transmitted by invertebrate vectors (i.e., mosquitoes, ticks, fleas, and other bloodsucking arthropods). Climate influences the behavior, survival, and reproduction rate of vectors, in turn affecting the suitability, distribution and abundance of habitats. Numerous studies have shown that the transmission patterns of diseases—such as tick-borne encephalitis, Rift Valley fever, Lyme disease, and West Nile fever—are strongly influenced by climatic conditions. In addition to invertebrate-transmitted zoonoses, direct-transmission zoonoses—such as hantavirus infections brought to humans by contact with wild rodents—can also be affected by climate change. In northern Europe, it has been observed that large human outbreaks of hemorrhagic fever caused by hantavirus—a disease with rodents as natural hosts—coincide with the growth peaks of rodent populations, favored by milder winter temperatures. In northern Italy, there has been a recent and sudden increase in seroprevalence for hantavirus in wild mice. Many zoonotic agents conveyed to human beings by water and food—such as *Salmonella* species, *Campylobacter jejuni*, *Escherichia coli*, and *Norovirus* species—show specific seasonal patterns of incidence. A study conducted on data from various European countries showed that cases of human salmonellosis increased by 5 to 10 percent for each degree of increase in the average weekly temperature.[4]

CLIMATE CHANGE, HUMAN HEALTH, AND THE HEALTH OF ECOSYSTEMS

Human health and the health of ecosystems are inextricably linked. The understanding of these interconnections between human health and the natural environment has increased rapidly in recent decades.[5] We must remember that *Homo sapiens* is one of the many species in the ecosystem, and our survival is linked to good ecological status. The weight of environmental risk factors on diseases is significant and ranges from air and water pollution to the impacts of UV radiation on skin cancer. Altered environments can cause up to one in four deaths globally.[6] Climate change has a series of impacts on ecosystems that affect human health. These impacts range from those relating to the availability of water to those on diseases transmitted by vectors on biodiversity,[7] to deterioration of the chemical and ecological quality of the environment. Green spaces are also likely to be affected by climate change in multiple ways, with consequent negative health effects such as the inability of ecosystems to regulate air quality and to mitigate heat waves caused by extreme temperatures.

Future scenarios envisage an increase in the world population to eight billion people by 2030, which could lead to serious food, water, and energy shortages, with severe health consequences and resource scarcity. The loss of the services provided by natural ecosystems will lead to the need to seek expensive alternatives. Investments in our natural capital will allow savings in the long term, and for this reason,

---

[4] See Jan C. Semenza and Jonathan E. Suk, "Vector-Borne Diseases and Climate Change: A European Perspective," *FEMS Microbiology Letters* 365, no. 2 (2018): fnx244.
[5] See Aline Chiabai et al., "The Nexus between Climate Change, Ecosystem Services and Human Health: Towards a Conceptual Framework," *Science of the Total Environment* 635 (2018): 1191–1204.
[6] See M. Neira and A. Prüss-Ustün, "Preventing Disease through Healthy Environments: A Global Assessment of the Environmental Burden of Disease Abstract," *Toxicology Letters* 259 (2016): S1.
[7] See Laura Mancini et al., "Global Environmental Changes and the Impact on Ecosystems and Human Health," *Energia, Ambiente e Innovazione* 3 (2017): 98–105; Anthony J. McMichael, Rosalie W. Woodruff, and Simon Hales, "Climate Change and Human Health: Present and Future Risks," *Lancet* 367, no. 9513 (2006): 859–869.

they are essential for our well-being and future survival.[8] Furthermore, ecosystem approaches for adaptation to climate change or adaptation based on the ecosystem involve a wide range of management activities to increase the resilience of ecosystems and reduce the vulnerability of people and the environment and, thus, the effects on human health and well-being.

Environmental health has always been a pragmatic discipline focused on identifying, quantifying, and addressing threats to human health in the environment. This ecosystemic vision reverses the way we deal with problems and allows us to implement preventive measures without having to constantly respond to emergencies. The analysis of effects induced by climate changes on the average values of temperature and rainfall can clarify their variation, which gives rise to extreme events such as floods, heat waves or cold spells. The sudden increase in rainfall can give rise to significant phenomena such as floods in urban areas, and thus represents a direct or indirect risk to health. When ecosystems are out of balance, such adverse events can have even greater negative impacts. While the effects of climate change on health can only be estimated, a WHO assessment, which considers a subset of possible health impacts, concluded that since 1970, warming has caused over 140,000 more deaths per year, while estimating an additional 250,000 deaths per year in the future.[9]

Health effects can be detected in a short-, medium- and long-term periods. All the effects induced by extreme events are recorded internationally in databases such as the International Disaster Database (EM-DAT). Recent studies on medium-term effects have shown an increase in the incidence of infectious diseases in the population correlated with the occurrence of extreme events.[10] Transmission of the causative agent can be linked directly or indirectly to water. The incidence of some of the diseases examined—such as hepatitis A, Legionnaires' disease, and infectious diarrhea—increased

---

[8] See European Commission, "Ecosystems Goods and Services," September 2009, ec.europa.eu/environment/pubs/pdf/factsheets/Eco-systems%20goods%20and%20Services/Ecosystem_EN.pdf.
[9] See World Health Organization, "Climate Change and Health."
[10] See Ankie Sterk et al., "Direct and Indirect Effects of Climate Change on the Risk of Infection by Water-Transmitted Pathogens," *Environmental Science & Technology* 47, no. 22 (2013): 12648–12660; Hans J. P. Marvin, Gijs A. Kleter, H.J. Van der Fels-Klerx, Maryvon Y. Noordham, Eelco Franz, Don J.M. Willems, and Alistair Boxall, "Proactive Systems for Early Warning of Potential Impacts of Natural Disasters on Food Safety: Climate-Change-Induced Extreme Events as Case in Point," *Food Control* 34, no. 2 (2013): 444–456; Lisa Brown and Virginia Murray, "Examining the Relationship between Infectious Diseases and Flooding in Europe: A Systematic Literature Review and Summary of Possible Public Health Interventions," *Disaster Health* 1, no. 2 (2013): 117–127; Emily K. Shuman, "Global Climate Change and Infectious Diseases," *New England Journal of Medicine* 362, no. 12 (2010): 1061–1063.

in the period immediately following flooding, linked to the incubation times of the infectious agents in cities suffering major floods. Harm to people is added to property damage, with cities among the most vulnerable areas.[11] Defense requires an ecosystem in balance. The interconnections between climate change, ecosystems, and health must be properly understood in order to plan better adaptive responses and ensure that potential health benefits can be taken into account in the design of adaptation measures, particularly with "nature-based solution" proposals.

Balanced and resilient ecosystems can cushion or reduce the effects on health. For example, good functionality of wetlands plays an important role in water resources because it increases the availability of water and its quality by reducing eutrophication processes, providing a cleaner environment. Moreover, good functionality of urban green spaces allows the preservation of air quality and the mitigation of heat waves caused by extreme temperatures.[12]

ACTION, RESTORATION, AND STRENGTHENING OF THE NATURAL AND ANTHROPIC ECOSYSTEM

It is fundamental to recognize, identify, inventory, and map multiple functions and services provided by ecosystems on different scales, as well as to link ecosystem management with sustainable livelihoods and development (i.e., by demonstrating clear social and economic benefits for investing in ecosystem management). Politically and socially, greater awareness of the economic importance of ecosystem goods and services must also be acquired. Indeed, the future commitment of the WHO is to move from the awareness, knowledge, and denunciation phase to global environmental economic policy actions. The goal is to reduce risks by implementing surveillance plans or communicating existing ones. Moreover, the necessary implementation of these plans for large-scale prevention involves the environmental and health policies of all countries. Public participation in prevention is fundamental, and the need to involve and educate citizens on risk and prevention is an important part of all the most recent regulations in this sector. In conclusion, it is important to recognize and highlight the signals that our planet is giving us, in particular our ecosystems which are closely connected with the well-being and health of populations. Signals of alteration or deterioration of the ecosystem should be considered an alarm by policymakers, so

---

[11] See Stefania Marcheggiani et al., "Risks of Water-Borne Disease Outbreaks after Extreme Events," *Toxicological & Environmental Chemistry* 92, no. 3 (2010): 593–599.

[12] See Millennium Ecosystem Assessment, *Ecosystems and Human Well-Being: Synthesis* (Washington, DC: Island Press, 2005).

they can apply preventive measures to protect human health. The WHO has emphasized the need for a new perspective focused on ecosystems and on the recognition that long-term health in human populations is mainly based on the stability and continuous functioning of life-support systems in the biosphere.

The future success of global human well-being and health now requires that we break down the interdisciplinary barriers that still separate human and veterinary medicine from ecological, evolutionary, and environmental sciences. The development of integrative approaches should be promoted by linking the study of factors underlying stress responses to their consequences on ecosystem functioning and evolution.[13]

Integrated strategies and holistic, transdisciplinary, and multisectoral approaches to health are pivotal to deal effectively and in a timely manner with the direct or indirect effects of climate change on health, now recognized by the scientific community as a most urgent public health problem. Therefore, the next ten years will prove crucial for the health of ecosystems, animals, humans, and the planet.[14] These considerations support the notion that health security, as a global good, must be understood on a global scale and from a global and crosscutting perspective, integrating human health, animal health, plant health, ecosystems health and biodiversity.

AVERTING CLIMATE BREAKDOWN BY RESTORING ECOSYSTEMS

International actions are emerging. The United Nations General Assembly declared 2021–2030 the UN Decade on Ecosystem Restoration. The degradation of land and marine ecosystems undermines the well-being of 3.2 billion people and costs about 10 percent of the annual global gross product in loss of species and ecosystems services. Key ecosystems that deliver numerous services essential to food and agriculture, including supply of freshwater, protection against hazards and provision of habitat for species such as fish and pollinators, are declining rapidly. The UN Decade on Ecosystem Restoration aims to massively scale up the restoration of degraded and destroyed ecosystems as proven measures to fight the climate crisis and enhance food security, water supply, and

---

[13] See Jan M. Baert et al., "Biodiversity Effects on Ecosystem Functioning Respond Unimodally to Environmental Stress," *Ecology Letters* 21, no. 8 (2018): 1191–1199; Food and Agriculture Organization of the United Nations, World Organization for Animal Health, and World Health Organization, *Taking a Multisectoral, One Health Approach: A Tripartite Guide to Addressing Zoonotic Diseases in Countries* (New York: Food and Agriculture Organization of the United Nations, World Organization for Animal Health, and World Health Organization, 2019).

[14] See Delphine Destoumieux-Garzón et al., "The One Health Concept: 10 Years Old and a Long Road Ahead," *Frontiers in Veterinary Science* 5, no. 14 (2018): doi: 10.3389/fvets.2018.00014.

biodiversity. Restoration could remove up to 26 gigatons of greenhouse gases from the atmosphere. The UN Environment Program and the Food and Agriculture Organization of the United Nations (FAO) will lead the implementation.

A recent estimate suggests that around one-third of the greenhouse gas mitigation required between now and 2030 can be provided by carbon drawdown through Natural Climate Solutions. Natural Climate Solutions, roughly speaking, mean ecological restoration. This implies drawing carbon dioxide out of the air by protecting and restoring ecosystems. By defending, restoring, and re-establishing forests, peatlands, mangroves, salt marshes, natural seabeds, and other crucial ecosystems, large amounts of carbon can be removed from the air and stored. At the same time, the protection and restoration of these ecosystems can help to minimize a sixth great extinction, while enhancing local people's resilience against climate disaster. Defending the living world and defending the climate are, in many cases, one and the same.

This approach should not be used as a substitute for the rapid and comprehensive decarbonization of industrial economies. A committed and well-funded program to address all the causes of climate chaos, including Natural Climate Solutions, could help us hold the heating of the planet below 1.5 °C. Of the ten types of co-benefit, the most frequently studied is the ecosystem impact.[15]

ROME INTERNATIONAL CHARTER ON HEALTH AND CLIMATE CHANGE

The 2019 Symposium "Health and Climate Change" promoted an intersectoral and multidisciplinary approach to estimate and to prevent climate change-related events, as well as to call on the authorities to put into place measures to reduce adverse health effects. Approximately five hundred researchers from twenty seven countries gathered at the Symposium to discuss the far-ranging impacts that climate changes have and will increasingly have on human health and to reach a consensus on a set of key actions required to face the risks and threats.

At the end of the Symposium, the Rome International Charter on Health and Climate Change was presented. The methodology moves from analysis to assessment of the most relevant scientific evidence related to the impact and consequences of climate change on human animal and environmental health to advocacy and key measures and

---

[15] See Hong-Mei Deng et al., "Co-Benefits of Greenhouse Gas Mitigation: A Review and Classification by Type, Mitigation Sector, and Geography," *Environmental Research Letters* 12, no. 12 (2017): doi: 10.1088/1748-9326/aa98d2.

messages.[16] The Charter includes a series of actions and recommendations, discussed, and shared by all the participants, intended to inform policymakers and all the stakeholders involved in the management of climate changes. These recommendations support the notion that, as a global good, health security must be understood on a global scale and from a global and crosscutting perspective, integrating human health, animal health, plant health, ecosystems health, and biodiversity.[17] Its key focuses concern information, education, and empowerment of citizens as effective means to contribute to the mitigation of the effects of climate change on health and, at the same time, enable adaptation by fostering resilience of individuals and communities.

These key messages and measures need to be translated into action by decision-makers and administrators. They should be taken up and included in all policies if we want to protect the future of humankind and the planet (see Table 1).[18]

CONCLUSION

The health community has a vital role to play in accelerating progress to tackle climate change. The unacceptably high and potentially catastrophic health effects of climate change are being felt today. Tackling climate change could be the greatest global health opportunity of the twenty-first century.[19] Achieving a decarbonized global economy and securing the public health benefits it offers is no longer primarily a technical or economic question. It is now a political one. Many mitigation and adaptation responses to climate change are "no regret" options. In this context, the following actions are needed to assure health outcomes: health adaptation in all sectors and spending; adaptation delivery and implementation; adaptation planning and assessment; climate information service for health; communicating about health and climate risks and opportunities; ensuring that mitigation strengthens public health; adapting to face new and emerging health risks; reducing emissions from health

---

[16] See Walter Ricciard et al., "Health and Climate Change: Science Calls for Global Action," *Annali dell'Istituto Superiore di Sanità* 55, no. 4 (2019): 323–329.

[17] See World Health Organization, *Connecting Global Priorities: Biodiversity and Human Health—A State of Knowledge Review* (Geneva: World Health Organization, 2015).

[18] See Ricciardi et al., "Health and Climate Change."

[19] See Nick Watts et al., "The 2019 Report of the *Lancet* Countdown on Health and Climate Change: Ensuring That the Health of a Child Born Today Is Not Defined by a Changing Climate," *Lancet* 394, no. 10211 (2019): 1836–1878; Ricciardi et al., "Health and Climate Change."

services; encouraging a transition to cities that support and promote healthy lifestyles (see Figure 1).[20]

Table 1. Rome International Charter on Health and Climate Change[21]

Health is climate-dependent, so is our future: **Measures and Messages**	
Environment and health	Adopt climate change mitigation measures to reduce the environmental burden of pollutants and related human diseases.
Climate change and zoonoses	Promote a one-health approach (animal-human-environment) in both research and management programs.
Climate changes scenario	Adopt vigorous adaptation and mitigation measures to reduce the role of climate change as a multiplier of stressors, accelerating conflicts and societal fragmentation.
Climate change and children health	Recognize parks and protected areas as a vital source of health and well-being and pivotal in reconnecting children to nature and in mitigating the effects of climate change.
Healthier cities	Connect science to politics and the population and provide health and climate change education for healthier cities.
Mental health and climate change	Monitor fragility and resilience both at mental and psychosocial level and promote interventions.
Blue and green space	Provide education about management of, and access to, blue and green spaces to sustain and improve the physical and mental health and well-being of individuals and communities, particularly to overcome socio-economic inequalities.

---

[20] See Watts et al., "The 2019 Report of the *Lancet* Countdown on Health and Climate Change."
[21] The Table is modified from Ricciardi et al., "Health and Climate Change," 327.

Water, sanitation, and climate change	Adopt a holistic approach in policy, research, and management to strengthen climate adaptation and the resilience of water and sanitation systems, based on risk analysis and through water and sanitation safety plan management approaches.
Communicable disease and climate change	Monitor epidemic precursors of disease (human infections, climatic-, environmental-, vector-, social-, animal-, food-related, etc.) through a continuum of surveillance across sectors for the early detection of unusual patterns and for the improvement of prevention and control.
Health and climate change. Joint action for sustainable development	Consider externalities (calculations and embedding into prices), goods, and services to reduce carbon hotspots, social and financial return on investment of sustainable health systems (i.e., sustainability beyond economic dimensions, including social and environmental aspects).
Air quality, low carbon policy, health, and climate change	Place protection and promotion of health at the center of the climate change agenda, ensuring that policies to accelerate progress towards the zero-carbon economy capitalize on the health and wider economic benefits and communicate those benefits to the public and policymakers. Develop well designed policies to reduce the emissions of carbon dioxide and short-lived climate pollutants in sectors such as transport, energy, housing, urban design, health care, food, and agriculture.
Ecosystem and health	Link ecosystem management with sustainable livelihoods and development to avoid the sixth mass extinction.
Global health and climate change	Interrupt the vicious circle between climate change and air pollution by addressing its impact on chronic non-communicable diseases and on cardiovascular and respiratory diseases in particular.
Tools and needs	Develop tools that inform people and engage them on the issues. Empower people to take action at a personal level to undertake targeted intervention both for their own health and the environment.

From the environment friendly green to the healthy hospital	Make smart green technology available in hospitals, health centers, and in health care systems to respond to disasters related to climate change. Reduce the impact of clinical effects to improve resilience of affected citizens and to advocate for a persistent better health of the global population.
Food security–food safety and climate change	Adopt holistic approaches to food security, food safety, and climate change. Promote interconnectivity and cooperation (i.e., connect-collaborate-co-design) to build resilience to the effects of climate changes, harnessing disruption rather than being passively subjected to it.
Stakeholders	Encourage mindful, responsible, and effective transfer of scientific data on climate change and its effects among stakeholders to promote global and local health and well-being and to reduce vulnerability through education, training, and information/communication.

Figure 1. Health outcomes

Walter Ricciardi, MD, is President of the World Federation of Public Health Associations. He graduated in medicine in 1986 and earned a doctorate in public health medicine in 1990 from the University of Naples. He currently holds the title of Professor of Hygiene and Public Health at the Università Cattolica del Sacro Cuore in Rome where he is also Director of the Department of Public Health and Deputy Head of the Faculty of Medicine. In addition to contributing to over three hundred academic papers—primarily in the fields of epidemiology, health services research, and public health genomics—he is also editor of the *European Journal of Public Health*, of the *Oxford Handbook of Public Health Practice*, and founding editor of the *Italian Journal of Public Health*. Professor Ricciardi was Chair of the Public Health Section of the Higher Health Council. In 2011, the Minister of Health of Italy appointed him as his representative in the State-Region Committee for the evaluation of the Italian National Health Service. Internationally, he is a member of the European Commission expert panel on "Investing in Health," a member of the National Board of Medical Examiners of the United

States of America, and was elected President of the European Public Health Association from 2010 to 2014.

Dr. Laura Mancini is a naturalist with more than twenty years of experience studying aquatic ecosystems and their interactions with human health. She is a senior researcher at the Italian National Institute of Health (Istituto Superiore di Sanità) and head of the Environmental Quality and Fish Farming Unit, leading the BlueHealth project, which studies populations living in proximity of waterways and coastal areas. She is also steering committee member of a research project in the Appia Antica Regional Park in Rome. Finally, she was previously involved in several large pan-European projects and currently is implementing the European funded Water Framework Directive 2000/60/CE in both Italy and Europe.

The Vital Contribution of Independent, Ethically Grounded Research to the Global Health Agenda

Kurt Straif

THE GLOBAL BURDEN OF CANCER IS HIGH and continues to increase. The annual number of new cases was estimated at 18 million in 2018 and is expected to increase to almost 30 million by 2040.[35] With current trends in demographics and exposure, much of the increase in cancer burden is expected in low- and medium-income countries. These countries often do not have the necessary resources to adequately treat all cancer patients. Cancer is already the leading underlying cause of death in many countries, and no country can treat its way out of the cancer problem. Therefore, prevention is the most effective response to this increasing challenge. The first step in cancer prevention is to identify the causes of human cancer.

THE IARC MONOGRAPHS PROGRAMME AND TRANSPARENCY

Through the Monographs Programme, the International Agency for Research on Cancer (IARC) seeks to identify the causes of cancer in the human environment.[36] The IARC Monographs Programme is an international, interdisciplinary approach to carcinogenic hazard identification. Soon after the Agency was established in 1965 as the specialized cancer research arm of World Health Organization (WHO), it received frequent requests for advice on the carcinogenic risk of chemicals, including inquiries about lists of known and suspected human carcinogens. In response to these requests, the IARC Monographs was initiated in 1971. Reviews and evaluations of nominated agents and exposures are carried out by Working Groups of scientific experts who are invited to participate on the basis of their expertise and their contributions to the relevant areas of science. Over

---

[35] See Freddie Bray et al., "Global Cancer Statistics 2018: GLOBOCAN Estimates of Incidence and Mortality Worldwide for 36 Cancers in 185 Countries," *CA: A Cancer Journal for Clinicians* 68, no. 6 (2018): 394–424.
[36] See International Agency for Research on Cancer (IARC), "Preamble to the IARC Monographs," last updated January 2019, monographs.iarc.fr/iarc-monographs-preamble-preamble-to-the-iarc-monographs/.

the decades, the IARC Monographs are a worldwide endeavor that has involved over a thousand scientists from more than fifty countries. The scientific reviews and evaluations are published as the *IARC Monographs on the Evaluation of Carcinogenic Risks to Humans*. In addition, and since volume 88 (2004), concise summaries of the evaluations and the rationale of each Monograph are published shortly after the meeting in the journal *Lancet Oncology*.

The Preamble to the IARC Monographs opens each volume.[37] The Preamble describes the objective and scope of the Programme, the scientific principles and procedures used in developing a Monograph, the types of evidence considered, and the scientific criteria that guide the evaluations. The Preamble promotes consistency of evaluations developed by different Working Groups on different topics. The evaluations are unique in representing the scientific consensus of an international Working Group of leading scientific experts. The carcinogenicity evaluation process used for the Monographs is widely recognized for its rigor and transparency (see Table 1). The Preamble also specifies strong measures to avoid conflicts of interest and interference by special interests.

Table 1: Measures to assure full transparency of the IARC Monograph Evaluations

Public announcements:	- Methods (published online, in advance)
	- Topic and timing (1 year in advance)
	- Working Group and all other participants (2 months in advance)
	- Results (*Lancet Oncology*, full *Monograph*)
Public process open to scientific observers	
Fully referenced *Monograph* is published online for free download:	- It cites only peer-reviewed, published, and publicly available data (for independent scientific scrutiny).
	- All studies (positive and negative) are described.
	- Rationale for conclusions is given.

---

[37] See International Agency for Research on Cancer (IARC), "Preamble to the IARC Monographs."

SELECTION OF PARTICIPANTS AND MANAGEMENT OF CONFLICT OF INTERESTS

The Monographs have a strong record of recruiting a diversity of international experts to form Working Groups that are authoritative, scientifically insightful, and highly regarded internationally. The Monographs identify leading experts from systematic searches of the literature and from public responses to the Call for Experts via the Monographs's website. Subject-matter experts are selected based on demonstrated knowledge and experience and careful screening to eliminate conflict of interests, and highly specific subject-matter expertise can be supplemented with broader subject expertise. Consideration is also given to demographic diversity and balance of scientific findings and interpretations. Each expert critically reviews an assigned portion of the scientific literature. The Working Group then meets at IARC for eight days to discuss the evidence, develop consensus, and arrive at an overall evaluation on the strength of the evidence.

The Monographs conduct open and transparent evaluations, strictly managing conflicts of interest of potential participants. Scientists with real or apparent conflicts of interest can be included as Invited Specialists. These scientists draft text on topics with no impact on carcinogenicity evaluation, but do not serve as meeting or subgroup chair and do not participate in the evaluation. Representatives from national and international public health institutions often participate in Monographs meetings because their agencies sponsor the Programme or are interested in the subject of the meeting. Stakeholders with potential conflicts of interest may be granted Observer status, in the interest of transparency, permitting them to observe the meeting but not influence the outcome. These clearly defined roles are an important mechanism by which subject-matter experts from different perspectives, including those who may have a real or apparent conflict, may contribute critical knowledge.

SYSTEMATIC REVIEW AND EVALUATION

The IARC Monographs are published as a series of volumes. Generally, three volumes are developed every year. Each volume contains one or more Monographs, which can cover either a single agent or a group of related agents. Each Monograph consists of a comprehensive, systematic review of the published scientific literature. Each Monograph contains a critical review of the openly available scientific literature, followed by an evaluation of the strength of the evidence that the agent can cause cancer. Over time, the structure of a Monograph has evolved to include the following sections: 1) Exposure data, 2) Studies of cancer in humans, 3) Studies of cancer in experimental animals, 4) Mechanistic and other relevant data, 5) Summary of sections 1-4, 6) Evaluation and rationale.

The process of systematic review and evaluation used for IARC's evaluations is laid out in the Preamble to the Monographs. International working groups of invited experts evaluate human, animal, and mechanistic evidence and reach a consensus evaluation of carcinogenicity for each agent. Initially, human and animal cancer data are evaluated separately according to causal criteria established in the Preamble. The strength of the evidence is categorized as Sufficient, Limited, Inadequate, or Suggesting lack of carcinogenicity. Mechanistic evidence can also be invoked to upgrade an evaluation. Alternatively, strong mechanistic evidence for the absence of a relevant mechanism in humans can downgrade an evaluation based on animal cancer data. For the overall evaluation of carcinogenicity, the Working Group considers the totality of the evidence and assigns agents to one of four causal groups following a well-defined evaluation matrix that evolved over time: 1) Carcinogenic to humans; 2A) Probably carcinogenic to humans; 2B) Possibly carcinogenic to humans; 3) Not classifiable as to its carcinogenicity to humans.

Consistent with the precautionary principle, sufficient animal cancer evidence can be used to classify an agent as potentially carcinogenic to humans in the absence of adequate human data. The evaluations undertaken for the Monographs constitute hazard identification, but the Preamble also aims at characterizing risk quantitatively.

Since its inception in 1971 the Monographs Programme has evaluated more than one thousand agents. Initially the Monographs focused on environmental and occupational exposures to chemicals, but their scope has expanded to include other factors and complex mixtures. Moreover, physical agents, biological agents, personal habits, and household exposures have been reviewed, some of them several times as new information became available in the published scientific literature. About 120 of these agents have been identified as carcinogenic, and about four hundred as probably carcinogenic, or possibly carcinogenic to humans (Groups 1, 2A and 2B, respectively).

VOLUME 100 OF THE IARC MONOGRAPHS

Volume 100 of the IARC Monographs comprises a reassessment and update of the more than one hundred agents classified by the IARC as carcinogenic to humans (Group 1) in Volumes 1-99. During 2008-2009, the IARC Monographs Programme organized six international Working Group meetings of experts in carcinogenesis and public health. A review article summarizes all evaluations of the volume 100 series, including cancer site-specific evidence (also covering agents currently evaluated as probably or possibly carcinogenic to humans), now allowing scientists to search the evidence on carcinogenic risks not only by agent, but also by cancer site.[38]

IMPACT OF THE IARC MONOGRAPHS

The international character of the IARC Monographs is unique. Moreover, as the specialized cancer agency of the World Health Organization, IARC has a strategic position within the United Nations. This status not only accords IARC a level of independence beyond that of most national agencies but elevates the impact of its work and facilitates the translation of Monograph findings to public health recommendations, guidelines, and policy worldwide.

The Monographs have been called the WHO's encyclopedia of carcinogens. National and international health agencies use the Monographs as a source of scientific information on known, probable, and possible carcinogens and as scientific support for their actions to prevent exposure to these agents. In high-income countries national agencies make frequent reference to the IARC Monographs. In developing countries, the IARC Monographs provide information about carcinogens to countries whose health ministries do not have the resources and experience in evaluating cancer hazards. Several Monographs have addressed cancer hazards that disproportionately affect developing countries, for example malaria (caused by infection with *Plasmodium falciparum* in holoendemic areas), betel quid chewing, and indoor emissions from household combustion of solid fuels (i.e., coal and biomass). In addition, as developing countries become more industrialized, exposure to some hazardous chemicals evaluated by the Monographs has increased, especially in countries where workplace and environmental standards are not adequate or are not enforced.

Individuals also use the information and conclusions from the Monographs to make better choices that reduce their exposure to potential carcinogens and their risk of developing cancer. In this way, the IARC Monographs contribute to cancer prevention and the improvement of public health.

SERIES OF IARC MONOGRAPHS ON AIR POLLUTION

In February 2003, the IARC Monographs Advisory Group on Priorities for Future Evaluations recommended that IARC develop a series of Monographs on air pollution. Emissions from motor vehicles, industrial processes, power generation, the household combustion of solid fuel, and other sources pollute the ambient air across the globe. The precise chemical and physical features of ambient air pollution, which comprise a myriad of individual chemical constituents, vary

---

[38] See Vincent James Cogliano et al., "Preventable Exposures Associated with Human Cancers," *Journal of the National Cancer Institute* 103, no. 24 (2011): 1827–1839.

around the world due to differences in the sources of pollution, climate, and meteorology, but the mixtures of ambient air pollution invariably contain specific chemicals known to be carcinogenic to humans.

Given the complexity of the topic, the multiplicity of environments where exposures to airborne carcinogens take place, the diversity of the sources, and the multiple components of the air pollution mixture that may contribute to its carcinogenicity, the 2003 Advisory Group recommended to convene a special Advisory Group for the planning of a series of IARC Monographs on air pollution.

A multidisciplinary Advisory Group[39] that included epidemiologists, toxicologists, atmospheric scientists, cancer biologists, and regulators first provided a state-of-the-art overview on topics related to exposure characterization, atmospheric and engineering sciences, epidemiological studies on cancer, results of pertinent cancer bioassays and data elucidating potential mechanisms of carcinogenicity of compounds related to air pollution, then discussed and made recommendations for the development of a series of Monographs on air pollution. Following these recommendations IARC developed a series of six Monographs meetings (see Table 2).

Table 2: Series of IARC Monographs on air pollution

Volume 92	Some Non-heterocyclic Polycyclic Aromatic Hydrocarbons and Some Related Exposures	October 2005
Volume 93	Carbon Black, Titanium Dioxide, and Talc	February 2006
Volume 95	Household Use of Solid Fuels and High-temperature Frying	October 2006
Volume 103	Bitumens and Bitumen Emissions, and Some N- and S-Heterocyclic Polycyclic Aromatic Hydrocarbons	October 2011
Volume 105	Diesel and Gasoline Engine Exhausts and Some Nitroarenes	June 2012
Volume 109	Ambient Air Pollution	October 2013

IARC MONOGRAPH ON DIESEL AND GASOLINE ENGINE EXHAUST

As part of this series of Monographs on air pollution, an IARC Monographs Working Group re-evaluated the carcinogenic hazards to humans of diesel and gasoline engine exhaust. Diesel engine exhaust

---

[39] See Kurt Straif et al., ed., *IARC Scientific Publication No. 161: Air Pollution and Cancer* (Lyon: International Agency for Research on Cancer, 2013).

was classified as "carcinogenic to humans" (Group 1) and gasoline engine exhaust as "possibly carcinogenic to humans" (Group 2B).[40]

Exposure is characterized in Section 1 of the Monograph. Diesel engines are used for on-road and non-road transport (e.g., trains and ships) and (heavy) equipment in various industrial sectors (e.g., mining and construction), and in electricity generators, particularly in developing countries. Emissions from these engines are complex with varying composition. The gas phase consists of carbon monoxide, nitrogen oxides, volatile organic compounds such as benzene and formaldehyde, and particles consist of elemental and organic carbon, ash, sulfate, and metals. Polycyclic aromatic hydrocarbons and nitroarenes are distributed over the gas and the particle phase.

Section 2 reviews the evidence on cancer in humans, typically derived from analytical studies of cancer epidemiology. A large US miners' study quantified diesel engine exhaust via estimated elemental carbon as a proxy of exposure and reported a cohort analysis and a nested case—control analysis adjusted for tobacco smoking. Both showed positive trends in lung cancer risk with increasing exposure, with a two- to three-fold increased risk in the highest categories of cumulative or average exposure. A 40 percent increased risk for lung cancer was observed in US railroad workers exposed to diesel engine exhaust. Indirect adjustment for smoking and a more accurate exposure assessment strengthened the validity of the results. A large cohort study reported a 15-40 percent increased lung cancer risk in US truck drivers and dockworkers with exposure to diesel engine exhaust, with a significant trend of increasing risks with longer duration of employment, adjusted for tobacco smoking. Findings of the above cohort studies were supported by those in other occupational groups and by case-control studies including various occupations involving exposure to diesel-engine exhaust supported the Working Group's conclusion of "sufficient evidence" in humans for the carcinogenicity of diesel engine exhaust.

Studies on cancer in experimental animals are summarized in Section 3. The Working Group concluded that there was "sufficient evidence" in experimental animals for the carcinogenicity of whole diesel engine exhaust, diesel engine exhaust particles, and extracts of the particles.

Section 4 summarizes evidence regarding mechanisms of carcinogenicity. Diesel engine exhaust, diesel engine exhaust particles, diesel engine exhaust condensates, and organic solvent extracts of diesel engine exhaust particles induced *in vitro* and *in vivo*

---

[40] See International Agency for Research on Cancer (IARC), *IARC Monographs on the Evaluation of Carcinogenic Risks to Humans. Volume 105: Diesel and Gasoline Engine Exhausts and Some Nitroarenes* (Lyon: International Agency for Research on Cancer, 2013).

various forms of DNA damage. Increased expression of genes involved in xenobiotic metabolism, oxidative stress, inflammation, antioxidant response, apoptosis, and cell cycle regulation in mammalian cells was observed. Positive genotoxicity biomarkers of exposure and effect were also observed in humans occupationally exposed to diesel engine exhaust. The Working Group concluded that there is "strong evidence" for the ability of whole diesel engine exhaust to induce cancer in humans through genotoxicity.

The Sections 5 and 6 summarize the evidence and conclude with the domain-specific and overall evaluations by the Working Group (see Table 3).

Table 3: Monograph evaluations on the carcinogenicity of diesel engine exhaust

• There is sufficient evidence for the carcinogenicity in humans of diesel engine exhaust. Diesel engine exhaust causes lung cancer. Also, a positive association between diesel engine exhaust and bladder cancer has been observed.
• There is sufficient evidence for the carcinogenicity in experimental animals of whole diesel engine exhaust.
• There is inadequate evidence in experimental animals for the carcinogenicity of gas-phase diesel engine exhaust.
• There is sufficient evidence in experimental animals for the carcinogenicity of diesel engine exhaust particulate matter. There is sufficient evidence in experimental animals for the carcinogenicity of extracts of diesel engine exhaust particles.
• Overall evaluation: Diesel engine exhaust is carcinogenic to humans (Group 1).

Gasoline engine exhaust and cancer risk was investigated in only a few epidemiological studies and, because of the difficulty to separate the effect of diesel engine exhaust and gasoline engine exhaust, evidence for carcinogenicity was evaluated as "inadequate." Organic extracts of gasoline engine exhaust condensate induced cancer in mice and in rats; gasoline engine exhaust condensate also induced cancer. The Working Group concluded that there was "sufficient evidence" in experimental animals for the carcinogenicity of condensates of gasoline engine exhaust. Overall, gasoline engine exhaust was classified as "possibly carcinogenic to humans" (Group 2B).

IARC MONOGRAPH ON GLYPHOSATE

The second example of IARC Monograph evaluations is on glyphosate. Its description here is more condensed and serves primarily as background for the concluding thoughts on interference from vested interest groups. Glyphosate was nominated for consideration by the 2014 Advisory Group (AG) to recommend priorities for IARC Monographs during 2015-2019, recommended for priority evaluation by that AG and included by the IARC Monographs scientific secretariat as one of the pesticides for evaluation by the Volume 112 Working Group.[41]

Glyphosate is currently the most heavily used broad-spectrum herbicide. It is used in more than 750 different products and its use has increased substantially with the marketing of genetically modified glyphosate-resistant crop varieties. In humans, there was limited evidence for the carcinogenicity of glyphosate. Case-control studies of occupational exposure in the US, Canada, and Sweden reported increased risks for non-Hodgkin's lymphoma that persisted after adjustment for other pesticides. The cohort of the US Agricultural Health Study did not show a significantly increased risk of non-Hodgkin's lymphoma. In experimental animals, in male CD-1 mice, glyphosate induced a positive trend in the incidence of renal tubule carcinoma, a rare tumor. Another study reported a positive trend for hemangiosarcoma in male mice. A glyphosate formulation promoted skin tumors in an initiation-promotion study in mice. Glyphosate has been detected in the blood and urine of agricultural workers, indicating absorption. Glyphosate and glyphosate formulations induced DNA and chromosomal damage in mammals, and in human and animal cells in vitro. One study reported increases in micronuclei in residents of several communities after spraying of glyphosate formulations. Bacterial mutagenesis tests were negative. Glyphosate, glyphosate formulations, and aminomethylphosphonic acid (AMPA, a metabolite) induced oxidative stress in rodents and *in vitro*.

Based on limited evidence in humans, sufficient evidence in experimental animals, and further corroborated by strong mechanistic evidence the Working Group classified glyphosate as "probably carcinogenic to humans" (Group 2A).

VESTED INTEREST

There is a long history of interference with science of occupational and environmental carcinogens by vested interest groups. Asbestos is probably the best-known example, and asbestos lobbyists followed

---

[41] See International Agency for Research on Cancer (IARC), *IARC Monographs on the Evaluation of Carcinogenic Risks to Humans. Volume 112: Some Organophosphate Insecticides and Herbicides* (Lyon: International Agency for Research on Cancer, 2017).

successful strategies developed by the tobacco industry: 1) creating doubt, 2) attacking science, 3) intimidating scientists, 4) lobbying policymakers, and 5) delaying action. Detailed case studies on global warming, second-hand smoke, asbestos, lead, plastics, and many other toxic materials are revealed in the thoroughly researched book by David Michaels, *Doubt Is Their Product: How Industry's Assault on Science Threatens Your Health*.[42]

Industry's reaction to the IARC Monograph on glyphosate is another stark case study. IARC and WHO were faced with an unprecedented, "orchestrated campaign" from industry. These reactions included:

- Demands to Directors of IARC and WHO to withdraw the *Monograph*'s evaluation
- Lobbying politicians, agencies, WHO, and member states
- Paid consultants criticize methods and findings
- Re-evaluation by an industry-funded committee
- Ghost-written scientific papers and press articles
- Legal demands aimed to harass US scientists
- Intimidating letters to international scientists
- Inquiries by the US Congress into the scientific rigor of the Monographs.

Activities from industry and IARC's careful, but firm, responses are documented on IARC's Governance website.[43] Further, several books—like *Whitewash: The Story of a Weed Killer, Cancer, and the Corruption of Science*[44]—and dossiers—like "The Monsanto Papers"[45]—are providing in depth details and background information, including from previously secret documents revealed during the discovery phase of ongoing litigation.

During these challenging times, the IARC Monographs have been strongly supported by the vast majority of the scientific community[46] and by civic society. With the increasing global burden of cancer and the urgent need for a stronger focus on prevention on one side and, on the other side, a general political climate of deniers of science in general and specifically of pollution and its nexus with the global

---

[42] See David Michaels, *Doubt Is Their Product: How Industry's Assault on Science Threatens Your Health* (Oxford: Oxford University Press, 2008).

[43] See International Agency for Research on Cancer (IARC), "Information for IARC Councils," governance.iarc.fr/ENG/infocouncils.php.

[44] See Carey Gillam, *Whitewash: The Story of a Weed Killer, Cancer, and the Corruption of Science* (Washington, DC: Island Press, 2017).

[45] See Stéphane Horel and Stéphane Foucart, "The Monsanto Papers," last modified November 20, 2017, *Environmental Health News*, www.ehn.org/monsanto-glyphosate-cancer-smear-campaign-2509710888.html.

[46] See Neil Pearce et al., "IARC Monographs: 40 Years of Evaluating Carcinogenic Hazards to Humans," *Environmental Health Perspectives* 123, no. 6 (2015): 507–514.

climate crisis and related massive adverse effects on human and planetary health, a strong and independent Monographs Programme is needed more than ever. Lessons learned from the exposed massive interference may help to strengthen authoritative and independent scientific consensus as the scientific basis for local, national, and global actions to prevent cancer. M

Kurt Straif, MD, is visiting scholar, Professor of Epidemiology, Co-Director of the Global Observatory on Pollution and Health of the Schiller Institute for Integrated Science and Society, at Boston College. After medical studies in Liège (Belgium), Heidelberg and Bonn (Germany), he started his career in internal medicine in Bonn. With a fellowship of the German Academic Exchange Office, he pursued an MPH at University of California Los Angeles (UCLA). Later, in Germany, he was assistant professor at the University of Giessen, focusing on occupational, environmental, and social medicine, and an assistant and associate professor in epidemiology and social medicine at the University of Münster. At the same time, he completed a PhD in epidemiology at the UCLA, with a focus on cancer epidemiology and epidemiologic methods. Since 2001, he had leading positions at the International Agency for Research on Cancer (IARC), first as a senior epidemiologist, Head of the IARC Monographs program, to classify carcinogenic hazards of all kinds of environmental exposures (chemical, biological, physical agents, and personal habits); Acting Head of a large epidemiological research group; and initiator of large international projects. In 2014, he relaunched the *IARC Handbooks of Cancer Prevention* with a broad perspective on prevention (breast cancer screening, avoidance of obesity, and colorectal cancer screening). Since 2017 he also supervises the World Health Organization Classification of Tumors ("Blue Books"). In 2016, he received the Champion of Environmental Health Research Award in commemoration of fifty years of Environmental Health Research by the National Institutes of Health. In 2018, he presented the Distinguished Lecture in Occupational and Environmental Cancer at the US National Cancer Institute.

Journal of Moral Theology

**Part 2:
The Changing Context of
Global Public Health**

Challenges Confronting Global Public Health

Keith Martin

AS OF THE WRITING OF THIS CHAPTER, COVID-19 has exploded across the globe infecting millions and killing hundreds of thousands of people. It has devastated the global economy, infusing many with a grim uncertainty about the future. The pandemic has been particularly lethal to those with pre-existing health conditions, the aged, and those living in low-resource communities. It has shone a bright, unsparing light on the structural disparities between and within countries that contribute to disproportionate suffering amongst the most vulnerable.

This threat should not have come as a surprise to anyone; we were forewarned. In 2019, as in previous years, the World Health Organization (WHO) listed pandemics as a top ten global health threat.[1] The Pentagon provided similar warnings in January 2017.[2] Despite this knowledge, the response to this novel coronavirus has varied from superb, in South Korea, New Zealand, Hong Kong, and Taiwan, to shambolic in the United States, Brazil, and the United Kingdom. The scale of human suffering and the immense economic toll this crisis has inflicted is due to the long-standing failure of many countries and the international community to invest appropriate resources into public health, to build the capacity and structures that can prevent and respond to lethal disease outbreaks, and to address many other challenges including those within the Sustainable Development Goals (SDGs).[3]

In 2015, the world came together and developed the Sustainable Development Goals. Its 17 goals and comprehensive list of 169 targets would, if achieved, put the world on a path to sustainable development

---

[1] See World Health Organization, "Ten Threats to Global Health in 2019," www.who.int/news-room/spotlight/ten-threats-to-global-health-in-2019.
[2] See Daniel R. Coats, "Worldwide Threat Assessment of the US Intelligence Community," *Global Sentinel*. Last modified February 2, 2018, globalsentinelng.com/2018/02/13/worldwide-threat-assessment-us-intelligence-community/.
[3] See World Bank, "Human Capital Project," www.worldbank.org/en/publication/human-capital; World Health Organization, "Sustainable Development Goals," www.who.int/sdg/en/.

to live on this planet, our only home, within its ability to provide the ecosystem services that constitute our life support systems. It was a significant leap from its predecessor the Millennium Development Goals (MDGs), which had a far narrower, biomedical focus on human health.[4] The SDGs include the health, environmental, legal, social and political determinants of health. They implicitly recognize that multidisciplinary and integrated approaches are needed to address the challenges we face. They acknowledged that the wellbeing of our species and that of our environment are inseparable. Today, five years after they were announced, most nations are far from achieving these goals prior to the 2030 deadline. Hence, how are we going to meet these targets?

Investing in global public health is central to achieving the SDGs because global public health is a platform that can address the triple threat we are facing: infectious diseases (both old and new); noncommunicable diseases (which account for 70 percent of mortality worldwide and are the leading cause of morbidity); and widespread environmental degradation, including climate change. Global public health calls for a range of interventions from vaccines, well-baby checkups, hypertension management, diabetes screening, nutritional supplementation, cancer screening, tobacco control, mental health screening, injury prevention, access to potable water, sanitation and others that can be delivered through public health platforms. In addition, public health can also be used to scale up low-cost environmental initiatives that reduce pollution, protect biodiversity, prevent disease spillover, mitigate climate change, preserve sources of freshwater, improve soil health, rehabilitate, and protect ecosystems.

The skills needed to scale up many of these interventions are generally not complicated to teach. The tools and infrastructure needed are relatively inexpensive. Despite its profound effectiveness, relevance, and relatively low cost, global public health is chronically underfunded.[5] This limit stems from a complex mix of political neglect, public inattention, poor advocacy, ineffective communication, and a fractured global public health community. The lack of attention amongst policymakers and the public is connected to a failure of the public health community to: mobilize the broad array of constituencies within public health; articulate clear, common messages showing the impact and value of public health; and being persistent in advocacy efforts. There is a failure to communicate the

---

[4] See World Health Organization, "Millennium Development Goals," (2020), www.who.int/topics/millennium_development_goals/en/.
[5] See Institute for Health Metrics and Evaluation, "Financing Global Health 2017," www.healthdata.org/infographic/financing-global-health-2017; "Foundation Funding in Global Health," *Health Affairs* 36, no. 12 (2017): 2207–2208.

rate of return for investing in global public health and the opportunity cost for failing to do so.

This chapter examines some of the challenges we are facing in global public health and solutions to address them. Crises present opportunities for change. COVID-19 is an opportunity to strengthen public health systems worldwide which will help to build a healthier, more secure, sustainable, equitable, and resilient future for all. Without doing so we will not achieve the SDGs and will be putting at risk the very future of our species.

INFECTIOUS DISEASES AND PANDEMIC PREVENTION

The 2014-2016 Ebola virus outbreak in West Africa shook the international community. Within two-and-a-half years, the virus killed more than 11,325 people and devastated the already fragile health systems in Guinea, Liberia, and Sierra Leone. In response to this crisis, the international community created the Global Health Security Agenda (GHSA)[6]—a platform for the international community to collaborate and prevent, detect, and respond to disease outbreaks. Today, 67 countries are signatories to it. The GHSA includes governments, international organizations, NGOs, faith-based groups, and the private sector. Gaps in this network are known. This awareness offers an extraordinary opportunity for countries to fill these gaps and expand the number of countries in the network. The broader and stronger this network is the less vulnerable everyone will be to a disease outbreak. As Dr. Tom Frieden, the former director of the Centers for Disease Control and Prevention used to say, "A weak health system anywhere is a weak health system everywhere."

The GHSA is a public health platform. Beyond monitoring and screening for emerging infectious diseases, it could also be used to scale up known public health programs to address existing pathogens including tuberculosis (i.e., the world's leading cause of death from infectious diseases), HIV, malaria, and neglected tropical diseases (NTDs). Public health interventions to address the NTDs have been exceptionally cost-effective investments. Successful examples are deworming schoolchildren, as well as the Carter Center's programs to eradicate Guinea worm (reducing the number of people afflicted from 3.5 million people in 1986 to 54 cases in 2019, a 99.99 percent reduction) and trachoma through surgery, antibiotics, facial cleanliness, and environmental improvement. Public health interventions can also be used to deploy one of the most cost effective, yet threatened interventions: vaccines.[7]

---

[6] See Global Health Security Agenda, "A Partnership against Global Health Threats," ghsagenda.org/.
[7] See Centers for Disease Control and Prevention, "Neglected Tropical Diseases," www.cdc.gov/globalhealth/ntd/index.html; The Carter Center, "Health Programs,"

Vaccine hesitancy is recognized by the WHO as a leading global health threat.[8] The anti-science and anti-vaxxer movements have grown stronger over time. In the United States and Europe, they have used the internet to spread fake conspiracies and false claims such as linking vaccines to autism, which has been thoroughly debunked. Notwithstanding, this movement has sown uncertainty in the public, which has resulted in sharp declines in vaccine coverage in various parts of the world. This decline has had lethal consequences. Measles outbreaks in 2019 in the US and in several European countries reached emergency levels.[9] Moreover, this decline is an obstacle to vaccinating all 11–12-year-olds against human papilloma virus (HPV). HPV infections contribute to the death of over 300,000 women/year from cervical cancer (most of whom live in low-income countries) and an increasing incidence in head and neck malignancies in men and women.[10] The scientific community needs to speak out against these false claims that are threatening people's lives.

NONCOMMUNICABLE DISEASES (NCDS)

NCDs include cardiovascular disease, cancer, respiratory diseases, and diabetes. (Mental health and injuries are often included as NCDs). NCDs are responsible for 41 million deaths a year or 71 percent of mortality worldwide. Over 75 percent of these deaths occur in low- and middle-income countries and nearly 33 percent of them are between 30 and 69 years of age. Eighty-five percent of these deaths in this age cohort are in low- and middle-income countries (LMICs). Thus, NCDs affect people at a younger age and behave more aggressively in LMICs than in high income countries.

Public health is central in preventing and treating these diseases. Tobacco use, physical inactivity, poor nutrition, the misuse of alcohol, and the abuse of drugs are all risk factors for acquiring an NCD. Several cost-effective public health measures to address NCDs are well-known. The WHO's "Best Buys and other Recommended Interventions for the Prevention and Control of NCDs," which updates the Global Action Plan for the Prevention and Control of NCDs, is a

---

www.cartercenter.org/; Shelly McNeil, "Overview of Vaccine Efficacy and Vaccine Effectiveness," www.who.int/influenza_vaccines_plan/resources/Session4_VEfficacy_VEffectiveness.PDF.

[8] See Editorial Board, "Vaccine Hesitancy: A Generation at Risk," *Lancet Child & Adolescent Health* 3, no. 5 (2019): 281.

[9] See Peter Hotez, "America and Europe's New Normal: The Return of Vaccine Preventable Diseases," *Pediatric Research* 85, no. 7 (2019): 912–914.

[10] See National Cancer Institute, "HPV and Cancer." Last modified January 22, 2021. www.cancer.gov/about-cancer/causes-prevention/risk/infectious-agents/hpv-and-cancer.

pragmatic list of high impact interventions that can save lives and reduce health costs.[11]

Another important opportunity to reduce the incidence of NCDs is to address the social determinants of health.[12] Reducing poverty, increasing personal safety and enabling people to access quality healthcare, education, housing, and a clean environment will have a profound impact on these diseases. Interventions such as Brazil's Bolsa Familia's conditional cash transfer program, started by President Luis Inácio Lula da Silva, reduced inequality, hunger, poverty, and disparities between urban and rural populations.[13]

All NCDs have a chronic component and therefore are associated with varying levels of disability. The personal and economic costs associated with NCDs are not articulated clearly or often enough when advocating for public health. Mental health is a prime example of this cost. Mental health is the leading cause of disability worldwide. There is a massive global shortage in access to mental health care services which is severest in low-income countries, poor communities, and refugee camps. According to the Global Health Observatory, in 2014-2016 low-income countries had 0.1 psychiatrists and 0.3 psychiatric nurses per 100,000 people. The rate of psychiatrists in high-income countries was 120 times greater, and for nurses was more than 75 times greater than in low-income countries. Liberia, which has been ravaged by war and the Ebola epidemic, has only two psychiatrists for its population of 4 million people. Simple screening and access to low-cost generic antidepressants and antipsychotics would have a profound impact on enabling patients to access essential mental health care. Due to the extreme shortage of mental health care professionals in most low-income countries, these services can be provided by trained public health providers.[14]

Another neglected area of NCDs in public health is injury and violence. In 2014, falls, road traffic accidents, industrial accidents,

---

[11] See World Health Organization, "Global Health Observatory Data: NCD Mortality and Morbidity," www.who.int/gho/ncd/mortality_morbidity/en/; Global Alliance for Chronic Diseases, "Home Page," www.gacd.org/; World Health Organization, "From Burden to Best Buys: Reducing the Economic Impact of Noncommunicable Diseases in Low and Middle-Income Countries," www.who.int/nmh/publications best_buys_summary.pdf.

[12] See Michael Marmot and Jessica J. Allen, "Social Determinants of Health Equity," *American Journal of Public Health* 104, Supplement 4 (2014): S517–S19.

[13] See Amie Shei et al., "The Impact of Brazil's Bolsa Familia Conditional Cash Transfer Program on Children's Health Care Utilization and Health Outcomes," *BMC International Health and Human Rights* 14 (2014): 10–14.

[14] See Elizabeth Reisinger Walker et al., "Mortality in Mental Disorders and Global Disease Burden Implications a Systematic Review and Meta-Analysis," *JAMA Psychiatry* 72, no. 4 (2015): 334–341; World Health Organization, "Global Health Observatory Data Repository: Human Resources Data by Country," apps.who.int/gho/data/node.main.MHHR?lang=en.

poisoning, drowning, burns, suicide, person-to-person violence, and other causes of injury caused more than five million deaths per year. This is 1.7 times the number of deaths due to HIV/AIDS, tuberculosis, and malaria combined. The impact on the families and loved ones of those who have died is incalculable. There are many known interventions that can prevent injuries and violence including: obligatory use of seatbelts, sidewalks, motorbike helmets, reducing speed limits, enforced blood alcohol limits for drivers, child resistant containers to prevent poisonings, modifying homes to prevent falls among the elderly, fencing pools to reduce drowning, educational programs to prevent intimate partner violence, and home visitations to reduce child maltreatment. Public health is designed to scale up these programs.[15]

ENVIRONMENTAL THREATS AND CLIMATE CHANGE

The last member of this triad of dangers is an existential threat to our survival and that of many species on the planet. We are destroying our environment at an unprecedented and unsustainable rate, massively overusing the ecosystem services nature provides that are essential for our survival: clean water, clean air, and nutrition. Nature is also a source of medicinal products, can prevent NCDs and infectious diseases, and provides protection from floods and inclement weather. A measure of our overuse of our environment is the Earth Overshoot Day. This marks the day when humanity's use of the planet's ecosystem services exceeds what Earth can regenerate in a year. In 2019 that day was July 29th. Twenty years prior to that, in 1999, it was September 29. This day moves earlier with every passing year.[16]

A neglected, yet critically important threat is pollution. The Lancet Commission on Pollution and Health clearly showed that pollution is the largest cause of disease and death in the world today causing an estimated nine million premature deaths annually. Despite its impact, pollution is an underreported contributor to the global burden of disease and affects people in low- and middle-income countries the hardest.[17]

Another significant, underappreciated, public health threat is our destruction of the planet's biodiversity through overuse, habitat destruction, pollution, climate change, and the growth of invasive

---

[15] See Spencer L. James et al., "Global Injury Morbidity and Mortality from 1990 to 2017: Results from the Global Burden of Disease Study 2017," *Injury Prevention*, 26 (2020): 96–114.
[16] See Earth Overshoot Day, "Home Page," www.overshootday.org/.
[17] See Philip J. Landrigan et al., "The *Lancet* Commission on Pollution and Health," *Lancet* 391, no. 10119 (2018): 462–512. See also Philip Landrigan's chapter in this book.

species. Anthropogenic activities have plunged the planet into a sixth extinction crisis. Why is this important? Wiping out species across the planet tears apart the foundation of the food chain upon which all life exists. This includes phytoplankton, which underpins life in the oceans. Its density is declining which is contributing to a significant reduction in the size and number of members of marine species living in the oceans. 1.5 billion people on the planet rely on the ocean for their food, not to mention their socioeconomic and cultural needs.[18] On land, an important backbone in the web of life—insects—is in rapid decline. Studies in Germany found that between 1989 in 2016 there was a shocking 76 percent drop in insect biomass. Insects are near the bottom of the food chain. Much of terrestrial life depends upon them. As the insects go, so too do we.[19]

Despite the international community pledging at the 2014 UN Climate Summit in New York to reduce deforestation by 50 percent over five years, a report released at the 2019 UN Climate Summit showed that it has surged 43 percent. The greatest loss is in the Amazon Basin. In Brazil alone, deforestation has increased 88 percent over this period. The Congo Basin and West Africa are also deforestation hotspots. Most of these forests and wetlands are being destroyed and converted to agricultural use. Eliminating these carbon sinks removes a natural mechanism to reduce atmospheric CO_2 and mitigate against climate change. This is also wiping out major areas of biodiversity losses and threatening the livelihoods and culture of the indigenous peoples who live in or near these areas.[20]

Our destruction of nature, as we are seeing with the COVID-19 pandemic, can create a global public health emergency. Over 70 percent of emerging infectious diseases are zoonotic. The SARS-CoV-2 that causes COVID-19 spilled over from a wild animal to humans. Our destruction of ecosystems is putting us into proximity to wild animals. Unsanitary wildlife markets are a major risk factor for spillover and were likely the cause of the current pandemic. Public health needs to have a much closer relationship with the veterinary science and environmental and conservation communities to advocate for: mainstreaming conservation into sustainable development initiatives; stopping demand for endangered species products;

---

[18] See World Health Organization, "Biodiversity and Health: The WHO-CBD Joint Work Programme," www.who.int/news-room/detail/01-01-2020-biodiversity-and-health-the-who-cbd-joint-work-programme.

[19] See Jeremy Hance, "The Great Insect Dying: Vanishing Act in Europe and North America," *Mongabay*. Last modified June 6, 2019. news.mongabay.com/2019/06/the-great-insect-dying-vanishing-act-in-europe-and-north-america/.

[20] See Fiona Harvey, "World Losing Area of Forest the Size of the UK Each Year, Report Finds," *The Guardian*. Last modified September 12, 2019, www.theguardian.com/environment/2019/sep/12/deforestation-world-losing-area-forest-size-of-uk-each-year-report-finds.

eliminating the trafficking in endangered species; incentivizing the public to change consumption habits; adopting new agricultural practices that can improve nutrition, reduce waste and increase output with a smaller environmental footprint; ending subsidies for the fossil fuel sector; supporting broader uptake of renewable energy sources and better use of freshwater resources; eliminating the clear cutting of forests; and strengthening emission and pollution standards.

Green spaces are also important in addressing noncommunicable diseases. Evidence shows that time spent in nature can have a profound impact upon mental health. These spaces provide opportunities for physical activity, which can reduce a person's risk for other NCDs.[21]

Climate change remains the ultimate existential threat we face. The International Panel on Climate Change has repeatedly warned that rising carbon dioxide emissions in the atmosphere due to anthropogenic activity will produce significant increases in global temperature, extreme weather events, and ocean warming and have significant impacts on food security, coastal flooding, biodiversity losses and more. This is a global crisis that poses a health, socioeconomic, and security threat to everyone.[22]

FINANCING, GOVERNANCE, AND CORRUPTION

We can ask for public health programs, but one must find a reliable source of funding to pay for them. The standard debate in international meetings usually revolves around increasing official development assistance (ODA). This shopworn discussion is measured by which countries will donate 0.7 percent of their GDP for ODA—the target set for high-income countries. This discussion however has little relationship to the resources needed to achieve the SDGs. ODA worldwide is approximately $147 billion a year, but the gap to achieve the sustainable development goals in developing countries alone is estimated to be about $2.5 trillion a year. Hence, where will the money come from?[23]

According to Transparent International, the world loses between $1.5 to $3 trillion per year to corruption. Weak enforcement, poor governance, ineffective public institutions in LMICs and high-income countries willing to turn a blind eye and be repositories of ill-gotten

---

[21] See Melanie Greaver Cordova, "Spending Time in Nature Reduces Stress," *Science Daily*. Last modified February 25, 2020. www.sciencedaily.com/releases/2020/02/200225164210.htm.

[22] See Intergovernmental Panel on Climate Change, "Summary for Policymakers of IPCC Special Report on Global Warming of 1.5°C Approved by Governments." Last modified October 8, 2018. www.ipcc.ch/2018/10/08/summary-for-policymakers-of-ipcc-special-report-on-global-warming-of-1-5c-approved-by-governments/.

[23] See United Nations, "Financing for the Sustainable Develeopment Goals," worldinvestmentforum.unctad.org/financing-for-the-sdgs/.

funds stolen from low-income countries allow this massive theft of funds from low-income nations. This scandal must change. Countries, especially high-income nations, must prosecute those involved in these activities, implement and enforce laws that stop money from being laundered through their economies and return stolen funds to countries from which they were taken. This intervention will inject resources that vastly overshadow any resources that can come from official development assistance and will largely meet the massive, existing funding gap in paying for the SDGs. As the WHO report "New Perspectives on Global Health Spending for Universal Health Coverage" showed, domestic public financing is the predominant source of health spending particularly for high and middle-income countries. For the least developed countries, domestic public financing represented only 22 percent of what was spent on health. Hence, for the least developed nations with the poorest health outcomes reducing corruption and returning stolen funds will make an enormous difference in their citizens' lives.[24]

Strengthening public health rests on having effective public institutions. The international community usually focuses on investing in specific diseases, other health-related challenges, vertical programs, and access to drugs and devices. While these investments are important, scant attention is spent on strengthening the foundation upon which sustainable development occurs. This foundation includes having a competent department of health, effective ministries of finance, justice, public works, environment, education, and independent, transparent oversight mechanisms for government operations and elections. This approach focuses on an integrated structure, where elected officials are accountable to the public—the basis of good governance. International development pays insufficient attention to strengthening public institutional capacity. This lack of attention needs to change since weak public institutions are a foundational disruptor of sustainable development. The strength, transparency, and effectiveness of public institutions is a primary determinant of state stability and thus of public health outcomes.

WOMEN'S HEALTH

Thankfully, women's health has received increased resources over the last twenty years. Significant successes have been achieved, but enormous gaps remain. According to UNICEF, although maternal

---

[24] See Stephen Johnson, "Corruption Is Costing the Global Economy $3.6 Trillion Dollars Every Year," *World Economic Forum*, www.weforum.org/agenda/2018/12/the-global-economy-loses-3-6-trillion-to-corruption-each-year-says-u-n/; World Health Organization, *New Perspectives on Global Health Spending for Universal Health Coverage* (Geneva: World Health Organization, 2018).

mortality declined by thirty-eight percent between 2000 and 2017, from 451,000 deaths in 2000 to 295,000 in 2017, this is still a staggering number of women dying from largely preventable complications in pregnancy and childbirth. For every woman who dies, approximately twenty others suffer serious injuries, infections, and disabilities. The geographic inequity is stark. Sub-Saharan Africa and South Asia account for eighty-six percent of maternal deaths worldwide.

In addition, in 2017, according to the Guttmacher Institute, 885 million women of reproductive age (fifteen to forty-nine years) in developing countries wanted to avoid pregnancy. Of these, 214 million cannot access modern contraception. Access to contraception is one of the most cost-effective ways of empowering women that will improve their health, economic, and social outcomes. Public health is an effective vehicle to provide prenatal and postnatal care and deliver the contraception options millions of women and men are seeking.[25]

NUTRITION

Our food systems have a significant impact upon everything from climate change to environmental degradation. Global population is currently at 7.7 billion and expected to reach 10.9 billion by the year 2100. Feeding this population will require significant increase in the availability of foodstuffs. With a growing demand for meat, this population increase has profound implications for land degradation and climate change. Addressing the population increase and its food demand will require a change in our consumption patterns. On top of this situation is the current burden of undernutrition and overnutrition. Overnutrition is a driver of NCDs, but undernutrition, particularly in childhood, is a risk factor for developing NCDs later in life. Public health is a platform that can influence consumption and address nutritional deficiencies.

CONCLUSION

Choose any SDG and you can easily identify a list of known, evidence-based solutions that can address them. Not only SDG number three (i.e., good health and well-being) but also the rest of the goals impact public health outcomes. The grand challenge today is not a lack of knowledge about how to address these problems; it is a lack of building capacity and willingness to implement what we already

---

[25] See World Health Organization, *Trends in Maternal Mortality 2000 to 2017: Estimates by WHO, UNICEF, UNFPA, World Bank Group and the United Nations Population Division* (Geneva: World Health Organization, 2019); Guttmacher Institute, "Adding It Up: Investing in Contraception and Maternal and Newborn Health, 2017," www.guttmacher.org/fact-sheet/adding-it-up-contraception-mnh-2017.

know will address them. This situation is rooted in a lack of human resources, infrastructure, reliable financing, and political will.[26]

Countries where public health outcomes are the worst share certain characteristics: weak public institutions and poor governance. This limitation cripples a nation's ability to manage a functional state which often leads to conflict, corruption, and poor socio-economic outcomes.

Scientists and the academic community understandably focus on the production of knowledge and the creation of papers for publication. This engagement is critically important to advance knowledge, but it is not enough. The production of knowledge must be matched by an even greater zeal to implement useful findings that come out of the research enterprise.

How global public health is practiced must also change. External actors, particularly those from high-resource environments, working in low-resource contexts, must be mindful that they are not doing something to others. Enormous power imbalances exist in these relationships which can lead to ineffective programming or, in worst case scenarios, cause harm. To avoid these pitfalls, local partners must be fully engaged from the beginning of any initiative, from its design right through to its implementation. Activities should not be done *to* others but *with* others. In addition, social and environmental impacts must be identified for any project. This attention is particularly important for vulnerable communities, which may be inadvertently left out in project planning but which can bear the brunt of a project's outcomes on their health, livelihood, and culture. In any development activity it is prudent to adhere to the medical maxim: do no harm.[27]

The challenges that we and our planet face today call for a significant change in how societies, nations, and the international community function. There is an urgency to this call to action for the consequences of the threats we are facing are profound. These challenges compel us to change our relationship with our environment, reform how societies are structured, change our consumption patterns, rethink incentives to affect change, and scale up what we already know will address these threats. To achieve this agenda we need to communicate, collaborate, and engage both policymakers and the public. One need not be partisan, but we must promote political action if we are going to address the challenges before us. What policies are enacted and where funds are allocated in the public sector are a matter of choice. We all need to influence those

---

[26] See International Finance Corporation, "Ethical Principles in Healthcare," www.ifc.org/wps/wcm/connect/ae698423-c213-4d28-8f18-a392cb45b6ff/EPIHC_draft_3-15-19_web.pdf?MOD=AJPERES.

[27] See World Bank, "Environmental and Social Framework," www.worldbank.org/en/projects-operations/environmental-and-social-framework; World Bank, "Human Capital Project."

choices for they will determine our fate and that of our planet as we know it.

Rudolph Virchow famously said, "Medicine is a social science, and politics is nothing but medicine on a large scale."[28] He recognized that social inequality is a root cause of ill health. We are certainly seeing a widening of disparities between and within countries. Biomedical and non-biomedical disciplines in global public health are all, to various extents, social sciences. Global public health is a window into societies' challenges. This awareness makes global public health practitioners ideally suited to be public scientists who create and advocate for the changes we desperately need. ∎

> "Bad men need nothing more to compass their ends, than that good men should look on and do nothing." - *John Stuart Mill, "Inaugural Address Delivered to the University of St. Andrews," February 1, 1867.*

Keith Martin, MD, is a physician who, since 2012, has served as the founding Executive Director of the Consortium of Universities for Global Health (CUGH) based in Washington, DC, which gathers over 170 academic institutions from around the world. Between 1993 and 2011, Dr. Martin served as a Member of Parliament in Canada's House of Commons and held shadow ministerial portfolios in foreign affairs, international development, and health. He also served as Canada's Parliamentary Secretary for Defense. In 2004, he was appointed to the Queen's Privy Council for Canada. His main areas of focus are global health, foreign policy, security, international development, conservation, and the environment. He is particularly interested in strengthening human resources capabilities and scaling up initiatives in low-income settings that improve environmental sustainability and human security. As a parliamentarian, Dr. Martin created *CanadaAid.ca*, an online platform to facilitate partnerships between universities, governments, multilateral institutions, NGOs, and the private sector. In 2006, Dr. Martin founded Canada's first all-party Conservation Caucus in Parliament and developed an online conservation site to help mainstream sustainable conservation and environmental practices. Dr. Martin has been on numerous diplomatic missions to areas in crisis including Sudan, Zimbabwe, Mali, Niger, Sierra Leone, Colombia, and the Middle East. He served as a physician in South Africa on the Mozambique border during that country's civil war. He has travelled widely in Africa. Dr. Martin is the author of more than 150 editorial pieces published in Canada's major newspapers and has appeared frequently as a political and social commentator on television and radio. He is currently a board member of the Jane Goodall Institute, editorial board member for the *Annals of Global Health*, and an advisor for the International Cancer Expert Corps. He contributed to the *Lancet* Commission on the Global Surgery Deficit, is a current commissioner on the *Lancet*-ISMMS

---

[28] Rudolf Virchow, "Der Armenarzt," *Medicinische Reform* 18 (1848): 125.

Commission on Pollution, Health and Development, and is a member of the Global Sepsis Alliance.

Journal of Moral Theology

Pollution, Climate Change, and Global Public Health: Social Justice and the Common Good

Philip J. Landrigan

POLLUTION AND CLIMATE CHANGE ARE TWO linked threats to planetary health, hallmarks of the Anthropocene epoch. Both threaten the health and well-being of all people and the sustainability of human societies.[1] Both are rooted in injustice and are inimical to the common good. Pollution and climate change are products of the linear, take-make-use-dispose economic model that Pope Francis has termed "the throwaway culture" (*Laudato Si'*, no. 16). In this paradigm, natural resources and human capital are viewed as abundant and expendable. Unwanted wastes are released into the environment with no concern for where they go or for whom they may harm. The consequences of reckless exploitation of the earth's resources are given little heed.[2] This economic model fails to link economic development to social justice or to stewardship of the planet.[3] It focuses single-mindedly on Gross Domestic Product (GDP),[4] is driven by a quest for endless growth, and is ultimately unsustainable.[5]

Pollution and climate change are closely linked.[6] Combustion of fossil fuel is the main cause of both. Fossil fuel combustion accounts for 85 percent of fine particulate air pollution and for almost all airborne emissions of sulfur oxides and nitrogen oxides. Fossil fuel combustion is the major source also of the greenhouse gases such as

---

[1] See Philip J. Landrigan et al., "The *Lancet* Commission on Pollution and Health," *Lancet* 391, no. 10119 (2018): 462–512.
[2] See Kate Raworth, *Doughnut Economics: Seven Ways to Think Like a 21st Century Economist* (White River Junction, VT: Chelsea Green, 2017); Sarah Whitmee et al., "Safeguarding Human Health in the Anthropocene Epoch: Report of the Rockefeller Foundation-*Lancet* Commission on Planetary Health," *Lancet* 386, no. 10007 (2015): 1973–2028.
[3] See Pope Francis, *Laudato Si'*; Whitmee et al., "Safeguarding Human Health"; Anthony J. McMichael et al., *Climate Change and the Health of Nations: Famines, Fevers, and the Fate of Populations* (New York: Oxford University Press, 2017).
[4] See Raworth, *Doughnut Economics*.
[5] See Johan Rockstrom et al., "A Safe Operating Space for Humanity," *Nature* 461, no. 7263 (2009): 472–475.
[6] See McMichael et al., *Climate Change and the Health of Nations*.

carbon dioxide, CO_2, and the short-lived climate pollutants that are the major drivers of climate change.[7] The impacts of pollution and climate change on human health and well-being are not evenly distributed. Instead, they fall disproportionately upon the poor and the disenfranchised, on women and children, on people in low-income and lower-middle-income countries, and on indigenous peoples around the world. This is environmental injustice on a global scale.[8] This chapter briefly summarizes current knowledge of the known and projected health effects of pollution and climate change and examines the distribution of their impacts through the lens of social justice.

POLLUTION

Pollution is the largest environmental cause of disease, disability, and death in the world today. In 2015, it was responsible for an estimated 9 million premature deaths, 16 percent of all deaths worldwide.[9] The number of deaths due to pollution is three times greater than the number due to AIDS, tuberculosis, and malaria combined, and ten times more than the number resulting from all wars and other forms of violence.

Air pollution accounts for the largest fraction of pollution-related diseases and premature deaths. In 2015, it was responsible for 6.4 million deaths: 2.8 million from household air pollution and 4.2 million from ambient air pollution. Water pollution caused 1.75 million deaths. Occupational pollutants caused 0.85 million deaths. Soil pollution, heavy metals, and toxic chemicals caused 0.5 million deaths.[10]

Chemical pollution is a growing global challenge. Its effects on human health are incompletely defined, and its impact on the global burden of disease is undercounted. An estimated 140,000 new chemicals and pesticides have been invented since 1950.[11] These are materials that never existed on earth. They are incorporated today into millions of consumer products, and many have become widely disseminated in the environment. National surveys conducted by organizations such as the US Centers for Disease Control and Prevention document the presence of several hundred manufactured

---

[7] See McMichael et al., *Climate Change and the Health of Nations*.
[8] See Landrigan et al., "The *Lancet* Commission on Pollution and Health"; Robert D. Bullard, *Dumping in Dixie: Race, Class, and Environmental Quality* (Boulder, CO: Westview Press, 1990).
[9] See Landrigan et al., "The *Lancet* Commission on Pollution and Health."
[10] See Landrigan et al., "The *Lancet* Commission on Pollution and Health."
[11] See Philip J. Landrigan, and Lynn R. Goldman, "Children's Vulnerability to Toxic Chemicals: A Challenge and Opportunity to Strengthen Health and Environmental Policy," *Health Affairs* 30, no. 5 (2011): 842–850.

chemicals in the bodies of most persons living on earth today.[12] Far too few of these materials have been tested for safety or toxicity and thus neither the full magnitude of their effects on human health nor their contribution to the global burden of disease can be accurately assessed.[13]

Ocean pollution is another worsening problem. It is a complex mixture of toxic metals, plastics, manufactured chemicals, petroleum wastes, industrial discharges, pesticides, pharmaceutical chemicals, agricultural run-off, and sewage. More than 80 percent arises from land-based sources. Ocean pollution has multiple adverse impacts on marine ecosystems as well as on human health. Petroleum pollutants[14] reduce expression of photosynthetic genes in *Prochlorococcus*, marine bacteria that produce 50-70 percent of the oxygen in the earth's atmosphere and sustain all terrestrial life.[15] Plastic pollution threatens marine mammals, fish, and seabirds.[16] Industrial releases, agricultural runoff, pharmaceutical waste, and sewage cause coastal pollution that increases the frequency of harmful algal blooms, worsens bacterial contamination of the seas, and contributes to the destruction of coral reefs.[17]

Ocean pollutants threaten human health. Health effects range from neurodevelopmental disorders among infants in the womb exposed to methylmercury and polychlorinated biphenyls (PCBs) through their mothers' unwitting consumption of contaminated fish;[18] brain damage and sudden death in persons exposed to powerful toxins produced by harmful algal blooms;[19] and severe wound infections and cholera in persons exposed to marine bacteria.[20] Ocean pollution and climate

---

[12] See Centers for Disease Control and Prevention, "National Biomonitoring Program," www.cdc.gov/biomonitoring/.
[13] See Landrigan, and Goldman, "Children's Vulnerability to Toxic Chemicals." See also Kurt Straif's chapter in this book.
[14] See UNESCO, "Facts and Figures on Marine Pollution," www.unesco.org/new/en/natural-sciences/ioc-oceans/focus-areas/rio-20-ocean/blueprint-for-the-future-we-want/marine-pollution/facts-and-figures-on-marine-pollution/.
[15] See Pedro Echeveste et al., "Decrease in the Abundance and Viability of Oceanic Phytoplankton Due to Trace Levels of Complex Mixtures of Organic Pollutants," *Chemosphere* 81, no. 2 (2010): 161–168.
[16] See Marcus Eriksen et al., "Plastic Pollution in the World's Oceans: More Than 5 Trillion Plastic Pieces Weighing over 250,000 Tons Afloat at Sea," *PLoS One* 9, no. 12 (2014): e111913.
[17] See UNESCO, "Facts and Figures on Marine Pollution."
[18] See Philippe Grandjean et al., "Cognitive Deficit in 7-Year-Old Children with Prenatal Exposure to Methylmercury," *Neurotoxicology & Teratology* 19, no. 6 (1997): 417–428.
[19] See Raphael M. Kudela et al., *Harmful Algal Blooms: A Scientific Summary for Policy Makers (IOC/Inf-1320)* (Paris: IOC/UNESCO, 2015).
[20] See Luis E. Escobar et al., "A Global Map of Suitability for Coastal Vibrio Cholerae under Current and Future Climate Conditions," *Acta Tropica* 149 (2015): 202–211.

change contribute to declines in fish stocks, and these declines threaten food security.[21]

Global patterns of pollution and pollution-related disease are changing. Deaths due to household air pollution and water pollution—the forms of pollution associated with profound poverty and traditional lifestyles—are declining in number.[22] These hopeful trends reflect years of work by United Nations agencies, faith communities, and NGOs to control household air and water pollution[23] as well as the introduction of new vaccines, antibiotics, and pediatric treatment protocols.

At the same time, however, ambient air pollution, chemical pollution, and soil pollution—the more modern forms of pollution—are on the rise with the largest increases seen in rapidly developing low-income and middle-income countries.[24] The forces responsible for recent increases in industrial, automotive, and chemical pollution include globalization, increasing demands for energy, growth of polluting industries, proliferation of toxic chemicals and pesticides, and the growing global use of cars, trucks, and buses. In the absence of aggressive intervention, the number of deaths due to ambient air pollution could double by 2050.[25]

More than 70 percent of the disease and premature death caused by pollution is due to non-communicable diseases.[26] Thus, pollution is responsible for 21 percent of all cardiovascular disease deaths worldwide, 26 percent of ischemic heart disease deaths, 23 percent of stroke deaths, 51 percent of deaths from chronic obstructive pulmonary disease, and 43 percent of deaths due to lung cancer. Pollution appears also to be linked to adverse reproductive outcomes, obesity, diabetes, and neurodegenerative diseases. Rates of these diseases are rising globally. In rapidly developing low-income and lower-middle-income countries, pollution is the single largest cause of death from non-communicable disease.[27]

---

[21] See Christopher D. Golden et al., "Fall in Fish Catch Threatens Human Health," *Nature* 534, no. 7607 (2016): 317–320.

[22] See Landrigan et al., "The *Lancet* Commission on Pollution and Health."

[23] See Jamie Bartram et al., "Global Monitoring of Water Supply and Sanitation: History, Methods and Future Challenges," *International Journal of Environmental Research and Public Health* 11, no. 8 (2014): 8137–8165; Gautam N. Yadama, *Fires, Fuel, & the Fate of 3 Billion: The State of the Energy Impoverished* (Oxford: Oxford University Press, 2013).

[24] See Landrigan et al., "The *Lancet* Commission on Pollution and Health."

[25] See Jos Lelieveld et al., "The Contribution of Outdoor Air Pollution Sources to Premature Mortality on a Global Scale," *Nature* 525, no. 7569 (2015): 367–371.

[26] See Landrigan et al., "The *Lancet* Commission on Pollution and Health."

[27] See Richard Fuller et al., "Pollution and Non-Communicable Disease: Time to End the Neglect," *Lancet Planetary Health* 2, no. 3 (2018): e96–e98.

CLIMATE CHANGE

The main driver of global climate change is a sharp increase in levels of carbon dioxide (CO_2) in the earth's atmosphere since the beginning of the Industrial Revolution.[28] For the 800,000 years before the present era, CO_2 levels in the atmosphere were relatively stable and fluctuated between 200 and 300 parts per million (ppm). However, in the past two centuries, and especially in the past 50 years, CO_2 levels have risen dramatically.[29] The current level is 413 ppm,[30] a CO_2 concentration not seen since dinosaurs walked the earth. The main driver of this increase has been the release into the atmosphere of enormous quantities of CO_2 from the steadily increasing combustion of fossil fuels—coal, oil, and gas.

CO_2 in the atmosphere acts as a heat insulator—a 'blanket'—around the earth. It traps the heat of the sun as well as heat generated by human activity. The result is that the average temperature of the surface of the earth has warmed by approximately one degree centigrade since 1880, and the rate of increase has accelerated since 1970. Sixteen of the world's seventeen warmest years have occurred since 2000.[31] This increase is not evenly distributed. In some parts of the world, temperature has increased little or not at all, but other places, especially the Polar Regions, have experienced increases as great as 2-3 degrees.[32]

Global climate change has numerous ecologic effects. It causes sea surface temperatures to rise and glaciers to melt. It increases the frequency and intensity of extreme weather events such as heat waves, heavy rainstorms, and hurricanes.[33] It increases the frequency and severity of droughts and wildfires. Absorption of increasing amounts of atmospheric CO_2 into the oceans causes acidification of the seas that destroys coral reefs and marine microorganisms.[34] Coral reef

---

[28] See Whitmee et al., "Safeguarding Human Health"; McMichael et al., *Climate Change and the Health of Nations*; Rockstrom et al., "A Safe Operating Space for Humanity"; Nick Watts et al., "The *Lancet* Countdown on Health and Climate Change: From 25 Years of Inaction to a Global Transformation for Public Health," *Lancet* 391, no. 10120 (2018): 581–630.

[29] See McMichael et al., *Climate Change and the Health of Nations*; Whitmee et al., "Safeguarding Human Health."

[30] See Scripps Institution of Oceanography, "The Keeling Curve," scripps.ucsd.edu/programs/keelingcurve/.

[31] See H.-O. Pörtner et al., ed., *IPCC Special Report on the Ocean and Cryosphere in a Changing Climate* (Geneva: IPCC Intergovernmental Panel on Climate Change, 2019).

[32] See Government of Canada, "Climate Observations in the Northwest Territories (1957-2012): Inuvik, Norman Wells, Yellowknife, Fort Smith," www.enr.gov.nt.ca/sites/enr/files/page_3_nwt-climate-observations_06-13-2015_vf_1_0.pdf.

[33] See Government of Canada, "Climate Observations."

[34] See Government of Canada, "Climate Observations."

destruction contributes to reductions in fish stocks. Sea surface warming and worsening marine pollution are expanding the geographic ranges of marine pathogenic bacteria. The result is that life-threatening bacteria such as *Vibrio* species, the bacteria that cause cholera, are moving poleward into cold, previously unpolluted waters.[35]

Climate change and pollution interact with each other. Heat waves are associated with increases in air pollution. Floods and coastal spills may cause chemical spills. Warmer, longer growing seasons result in increased use of pesticides. Rapidly rising temperatures in the circumpolar regions liberate legacy pollutants from ice and permafrost and alter the global distribution of chemical pollutants.[36]

Climate change has multiple adverse effects on human health, and these effects will become more prevalent in future years if climate change continues to worsen.[37] Heat waves cause deaths from heat shock and dehydration, especially in vulnerable populations such as young children and the elderly. Rising sea levels, floods, hurricanes, and wildfires lead to loss of life. Droughts cause crop failure, which can lead in turn to malnutrition, forced migration, and even war. Rising temperatures expand the geographic ranges and lengthen the breeding seasons of mosquitoes and other insect vectors of disease; increases in diseases such as malaria and dengue are the consequences. Rising sea surface temperatures will result in increases in the frequency of cholera outbreaks, and cholera is expected in coming years to move into new, previously unaffected areas.[38]

DISPROPORTIONATE IMPACTS OF POLLUTION AND CLIMATE CHANGE ON VULNERABLE POPULATIONS

Both pollution and climate change disproportionately kill the poor and the vulnerable. High-income countries have solved many of their worst pollution problems. They have established pollution control programs based on law, policy, technology, and regulations that are sustainably funded and backed by enforcement. In consequence, their air is relatively clean, their water is mostly safe to drink, and rates of pollution-related disease are low.[39]

In the low-income and middle-income countries of the Global South, by contrast, pollution is severe and in many places worsening. The world's highest levels of ambient air pollution are encountered in the rapidly expanding cities of low-income and lower-middle-income countries. Household air pollution and contamination of drinking

---

[35] See Escobar et al., "A Global Map."
[36] See Landrigan et al., "The *Lancet* Commission on Pollution and Health."
[37] See Government of Canada, "Climate Observations."
[38] See Escobar et al., "A Global Map."
[39] See Landrigan et al., "The *Lancet* Commission on Pollution and Health."

water are also heavily concentrated in low-income countries in the Global South.[40]

The result of this inequitable pattern is that people in low-income and lower-middle-income countries suffer disproportionately from disease, disability, and premature death caused by pollution. Nearly 92 percent of all pollution-related deaths occur in these countries, with the greatest numbers seen in previously poor, now rapidly industrializing nations.[41] In the most severely affected countries, pollution is responsible for more than one death in four. In countries at every level of income, disease caused by pollution is concentrated among the poor and the powerless, minorities, and the marginalized.

Pollution, poverty, poor health, and social injustice are closely intertwined. Thus, poverty contributes to ill health and causes disease and death by forcing people to live without clean water or adequate sanitation, or to live near polluting factories or downstream from hazardous waste sites. Pollution contributes to the intergenerational perpetuation of poverty by causing disease, disability, lost income, and increased healthcare costs among poor families that are unable to sustain these burdens. Families are thus locked into debt and poverty for generations to come. Additionally, poverty can attract pollution, as is seen in the behavior of unscrupulous corporations that time and again have deliberately located polluting factories and hazardous waste sites in poor, politically powerless communities, a phenomenon termed 'environmental injustice.'[42]

Children in all societies are at particularly high risk of pollution-related disease. Children's vulnerability reflects their disproportionately heavy exposure to pollution coupled with their biological sensitivity.[43] In 2016, pollution was responsible for 940,000 deaths in children worldwide, two-thirds of them in children under the age of five years.[44] The overwhelming majority of these deaths occurred among children in low- and middle-income countries. Most were due to respiratory and gastrointestinal diseases caused by polluted air and water. Pollution is linked additionally to multiple non-communicable diseases in children, including low birth weight, asthma, cancer, and neurodevelopmental disorders. Chemical pollutants are known to cause many of these diseases, but the full impact of chemical pollution on children's health is not yet known

---

[40] See Landrigan et al., "The *Lancet* Commission on Pollution and Health."
[41] See Landrigan et al., "The *Lancet* Commission on Pollution and Health."
[42] See Bullard, *Dumping in Dixie*.
[43] See Committee on Pesticides in the Diets of Infants and Children National Research Council, *Pesticides in the Diets of Infants and Children* (Washington, DC: National Academy Press, 1993).
[44] See Philip J. Landrigan et al., "Pollution and Children's Health," *Science of the Total Environment* 650, Part 2 (2019): 2389–2394.

because the potential toxicity of many chemical pollutants has not been characterized.[45]

The health impacts of climate change are also disproportionally severe in the Global South and in poor communities worldwide. Low-income and middle-income countries lack the resources to protect themselves against climate change and to rebuild their societies in the aftermath of climate-related disasters. This inequity was vividly displayed in Puerto Rico in the wake of Hurricane Maria in September 2017.

Poor countries face a heightened risk of cholera due to climate change. With growing ocean pollution and sea surface warming, *Vibrio cholera*, the bacterium that causes cholera, is spreading into new previously uncontaminated areas and exposing previously unexposed populations.[46] High-income countries will be able to protect their citizens against cholera through a combination of efficient water treatment and effective medical care. By contrast, risk of cholera outbreaks will be great in places in the Global South where the public health and health care systems are weak, and hygiene and sanitation systems are dysfunctional due to civil unrest, conflicts, and natural disasters.

Poor countries face the disproportionate impacts of declining fish stocks due to climate change.[47] These declines inequitably threaten food security and increase risk of malnutrition among people in small island nations, in the high Arctic, and in coastal fishing communities—vulnerable populations who depend heavily on fish for food and whose societal survival depends on the health of the seas.[48]

Children in poor countries are highly vulnerable to climate change. Components of climate change that will directly affect children's health include increased temperatures, increasing frequency and severity of weather extremes, and sea level rise. Effects will be more serious among children in low-income families in the world's poorest countries. The effects of climate change on children's health may be expected to become greater in the years ahead as the world becomes warmer, sea levels rise, geographic ranges of insect vectors carrying malaria and dengue expand, and changing weather patterns cause increasingly severe storms, droughts, and malnutrition, which, in turn, lead to forced migration and war.

Conclusion

Pollution and climate change are both the consequences of human behavior which can be solved by changes in behavior. The substantial

---

[45] See Landrigan, and Goldman, "Children's Vulnerability to Toxic Chemicals."
[46] See Escobar et al., "A Global Map."
[47] See Golden et al., "Fall in Fish."
[48] See Pörtner et al., *IPCC Special Report on the Ocean*.

successes achieved to date against air, water, and soil pollution in high-income countries and in some middle-income countries show clearly that pollution can be controlled and prevented. These successes demonstrate that pollution prevention improves health, reduces disease, extends healthy life and is highly cost-effective.[49] In the United States, every $1 USD invested in control of air pollution since 1970 has returned an estimated economic benefit of $30 USD (range of estimate, $4–88 USD).[50] Similarly, the removal of lead from gasoline in countries around the world has boosted economies by increasing the intelligence of billions of children who have come of age in relatively lead-free environments and who are thus more intelligent and more economically productive.[51]

Similarly, the reductions in CO_2 emissions achieved in many states and countries through mandated reduction in fossil fuel combustion and concomitant uptake of clean, renewable energy show that climate change can be mitigated. These successes show further that mitigation is politically and economically feasible. Thus, in July 2019, New York State enacted comprehensive energy and climate legislation and pledged to reduce greenhouse gas emissions by 85 percent by 2050. To meet this target, New York is developing the United States's largest offshore wind farm and collaborating with scientists and engineers in Ireland and Denmark to improve its electric power grid. New York has also created economic incentives for clean vehicles, and tax incentives for energy conservation. Idaho Power, the largest electric utility in a deeply conservative state, has pledged to produce 100 percent of its electricity from renewable sources by 2045. The United Kingdom has committed to net zero carbon emissions by 2050. New York, Idaho, and the United Kingdom are reaping health and economic benefits from these enlightened actions; they are creating new, high-paying jobs in the wind and solar energy industries; and they are reducing inequities in health and well-being caused by pollution and climate change.[52]

Sustainable, long-term control of pollution and mitigation of climate will require enlightened and humane leadership. It will demand that societies at every level of income emulate the bold interventions taken to date. It will challenge societies and their leaders

---

[49] See Landrigan et al., "The *Lancet* Commission on Pollution and Health."
[50] See Office of Air and Radiation US Environmental Protection Agency, *The Benefits and Costs of the Clean Air Act from 1990 to 2020* (Washington, DC: Environmental Protection Agency, 2011), www.epa.gov/sites/production/files/2015-07/documents/summaryreport.pdf.
[51] See Scott D. Grosse et al., "Economic Gains Resulting from the Reduction in Children's Exposure to Lead in the United States," *Environmental Health Perspectives* 110, no. 6 (2002): 563–569.
[52] See Philip J. Landrigan et al., "The False Promise of Natural Gas," *New England Journal of Medicine* 382, no. 2 (2020): 104–107.

to confront powerful vested interests, combat corruption, and put in place programs that reduce inequity, advance the common good, and promote social and economic justice. Societies will need to fundamentally change their patterns of production, consumption, and transportation. They will need to move away from the linear, GDP-driven economic paradigm now so widely prevalent. They must adopt a new social compact that is based on justice, rooted in the concept of the circular economy,[53] and guided by concern for the common good.[54]

Philip J. Landrigan, MD, MSc, FAAP, is Director of the Global Public Health and the Common Good program and director of the Global Observatory on Pollution and Health at Boston College. He is a pediatrician, public health physician, and epidemiologist. Author of over seven hundred scientific publications and ten books, in his research he uses the tools of epidemiology to elucidate connections between toxic chemicals and human health, especially the health of infants and children. He is particularly interested in understanding how toxic chemicals injure the developing brains and nervous systems of children and in translating this knowledge into public policy to protect health. In New York City, he worked for many years in the Icahn School of Medicine at Mount Sinai and he was involved in the medical and epidemiologic follow-up of twenty thousand 9/11 rescue workers. From 2015 to 2017, he co-chaired the *Lancet* Commission on Pollution and Health.

---

[53] See World Economic Forum, "Towards the Circular Economy: Accelerating the Scale-up across Global Supply Chains," www3.weforum.org/docs/WEF_ENV_TowardsCircularEconomy_Report_2014.pdf.

[54] See Francis, "Laudato Si'", nos. 18, 23, 54, 129, 35, 56–59, 69, 77–78, 84, 89, 96, 98, 201, 204, 225, 231–232.

Journal of Moral Theology

Global Public Health and Catholic Insights: Collaboration on Enduring Challenges

Michael D. Rozier, SJ

IN THE MID-TWENTIETH CENTURY, the field of medicine had to grapple with a single person, the physician, serving as both caregiver and scientist. The often-competing interests of patient care and human subject research created ethical conflicts that threatened to undermine the profession itself.[1] At its core, this was not a technical challenge but a conceptual one. More recently, the humanities and social sciences have had to contend with the upending of the traditional canons in literature, history, philosophy, and more.[2] Critical theory, for example, has helped lay bare some fundamental questions of disciplinary identity.[3] Again, this is a conceptual challenge to disciplines and one that, if ignored, would only serve to undermine them in the long run. Global public health, this essay suggests, has some underlying questions related to its identity and practice that, if answered, will only increase its relevance in the decades to come. These underlying challenges have technical components but are truly about the nature of the profession itself.

In some ways, the need to address such questions emerges from the fact that the context of global public health is rapidly changing. The rise of climate change[4] and the ever growing, long-term displacement of large populations[5] create issues that were not pressing when global public health first emerged. In recent decades, global health professionals have also had to face the fact that the good intentions of many actors in the Global North over the past several decades have

---

[1] See Henry K. Beecher, "Ethics and Clinical Research," *New England Journal of Medicine* 274, no. 24 (1966): 1354–1360.
[2] See Cornel West, "Minority Discourse and the Pitfalls of Canon Formation," *Yale Journal of Criticism* 1, no. 1 (1987): 193–201.
[3] See Ben Agger, *Cultural Studies as Critical Theory* (London: Falmer Press, 1992).
[4] See Nick Watts et al., "The *Lancet* Countdown on Health and Climate Change: From 25 Years of Inaction to a Global Transformation for Public Health," *Lancet* 391, no. 10120 (2018): 581–630.
[5] See Claire E. Brolan et al., "The Right to Health of Non-Nationals and Displaced Persons in the Sustainable Development Goals Era: Challenges for Equity in Universal Health Care," *International Journal for Equity in Health* 16, no. 1 (2017): doi.org/10.1186/s12939-016-0500-z.

maintained a power dynamic over the Global South.[6] This inability to truly address the colonialism of early international aid has almost certainly slowed true progress in building low-income health systems. New technologies such as persistent geo-location and telehealth present new opportunities for advancing health,[7] but they also bring similar challenges that come with technology, such as use by malicious actors. Global public health's success will depend on responding to this changing context and doing so in a way that helps better define the field itself.

Ultimately, the questions are unlikely to be answered by global public health alone. Instead, global public health can draw upon the insights of other disciplines to better understand the questions it faces. To that end, this essay offers how one global institution, the Catholic Church, might better contribute to the goals of global public health. More specifically, this essay asks two questions: How can the Catholic Church use its theological resources to address some enduring challenges in global public health? In so doing, how might that engagement also reshape the Church?

Elsewhere, it was suggested why the Catholic Church dialogues less with the field of public health than one might suspect.[8] One of the most obvious reasons is that the core motivation for Christian involvement in health care is continuing the healing ministry of Jesus and the stories of Jesus's healings from the Gospels largely show him caring for the acute illness of individuals. Therefore, the health care infrastructure of the Church was built largely to provide medical care to those who are ill. This dependence then creates a situation where investments of the past—material, financial, and intellectual resources—shape where one is willing to invest in the future. In addition, the Church is not unlike the rest of the world, which places greater priority on rescuing those who are sick today instead of considering whether limited resources are better used to prevent future illness.

There may be skepticism as to whether global public health and religious communities make good partners. Global health, adopting the position of its parent discipline of public health, seems most interested in engaging with faith communities when these communities provide logistical support.[9] For example, global health organizations are often happy to use churches or mosques as locations

---

[6] See Jin Un Kim et al., "A Time for New North-South Relationships in Global Health," *International Journal of General Medicine* 10 (2017): 401–408.

[7] See Damien Dietrich et al., "Applications of Space Technologies to Global Health: Scoping Review," *Journal of Medical Internet Research* 20, no. 6 (2018): e230.

[8] See Michael Rozier, "Religion and Public Health: Moral Tradition as Both Problem and Solution," *Journal of Religion and Health* 56, no. 3 (2017): 1052–1063.

[9] See Nathan Grills, "The Paradox of Multilateral Organizations Engaging with Faith-Based Organizations," *Global Governance* 15, no. 4 (2009): 505–520.

for health events or to use religious leaders as advocates for certain health campaigns. However, concern about more substantial engagement arises because of the situations where religious communities have moral positions that are contrary to the broad goals of public health.[10] These situations most often occur in areas of reproductive health or gender norms, such as religious opposition to contraception or religious support of child marriage. The questions about deeper engagement are often shared by the religious organizations themselves, who believe they will be made to compromise their values to collaborate. Public health carries the power of the state, and there is often an uneasy alliance between religious communities and the states in which they are located. It is therefore easier simply to avoid potential conflicts. These concerns are not insignificant, but they have likely played an outsized role in the potential relationship between global public health and religious organizations. While acknowledging very real barriers, this essay suggests that the Catholic Church is a natural ally for global public health efforts not just on matters of logistics but also on underlying issues related to the discipline of global public health itself.

CATHOLIC CONTRIBUTION TO GLOBAL PUBLIC HEALTH

There are several long-standing challenges in the field of global public health, which are likely best solved with intellectual resources found outside of the field of public health itself. One advantage is that public health is already a multidisciplinary field, accustomed to integrating knowledge from several disciplines to solve complex problems. Therefore, the idea of gleaning insights from religion and theology should be in keeping with public health's natural instincts. To that end, I offer three specific challenges that global public health must face more directly and suggest ways in which the Catholic Church might be able to contribute to a solution.

Vocation and Joy

A question whose time has come is what it means to have a vocation to work within the discipline of public health. Those in clinical care have long interwoven their sense of profession and vocation.[11] Physicians and nurses have a deep well from which to draw when they seek clarity as to their purpose in this world, and evidence shows that a personal sense of vocation confers many benefits,

---

[10] See Andrew Tomkins et al., "Controversies in Faith and Health Care," *Lancet* 386, no. 10005 (2015): 1776–1785.

[11] See Alan B. Astrow, "Is Medicine a Spiritual Vocation?" *Society* 50, no. 2 (2013): 101–105; Abraham M. Nussbaum, *The Finest Traditions of My Calling: One Physician's Search for the Renewal of Medicine* (New Haven, CT: Yale University Press, 2016).

including a reduced likelihood of burnout.[12] Understanding one's vocation gives the clinical professions a sense of coherence and a degree of companionship among those who work in clinical care. Those who work in global public health—epidemiologists, behavioral specialists, administrators, and environmental scientists among others—do not have as robust a sense of vocation. This deficit may in part be due to that fact that public health or global health is new compared to the healing professions. Nevertheless, there also does not appear to be much investment from the discipline in cultivating what it means to have a vocation to this work. However, the resources of the Church can be brought to bear on this important issue.[13]

Fostering vocation within global public health does not occur without stepping back from solving technical problems and looking more deeply at what makes us human. Although vocation is often narrowed to a calling to ordained ministry, within the Church it has a much broader understanding. More properly, it is recognizing one's gifts and considering how one might most effectively use those gifts. Within the Catholic Church, vocation obviously is rooted in God, both in terms of giving the gifts one possesses and in revealing where those gifts might best be used. Some professions have been more intentional about incorporating this conversation into professional training than others, but global public health is among those that rarely make time for this question. This is certainly to the detriment of the field and those who labor with such commitment within it.

If the field takes seriously the question of vocation, other benefits will likely follow. For example, some definitions of one's call ask one to consider whether what one does brings them joy. Yet how much does one hear about joy in the work of global health? Or one might discover that considering one's vocation reveals goals that go beyond technical expertise, as important as that is. Giving time to the question of vocation might reveal how important virtues, such as compassion or courage, are to one's life. Professionals would likely find greater personal satisfaction in a field that integrates professional expertise and vocational meaning. Even more, many of these virtues build more effective health systems. Evidence shows that patients are willing to travel further and pay more when they perceive their provider is

---

[12] See Andrew J. Jager et al., "Association between Physician Burnout and Identification with Medicine as a Calling," *Mayo Clinic Proceedings* 92, no. 3 (2017): 415–422; John D. Yoon et al., "The Association between a Sense of Calling and Physician Well-Being: A National Study of Primary Care Physicians and Psychiatrists," *Academic Psychiatry* 41, no. 2 (2017): 167–173.

[13] See Max Stackhouse, "Vocation," in *The Oxford Handbook of Theological Ethics*, ed. Gilbert Meilaender and William Werpehowski (Oxford: Oxford University Press, 2007), 189–204; Edward P. Hahnenberg, *Awakening Vocation: A Theology of Christian Call* (Collegeville, MN: Liturgical Press, 2010).

compassionate.[14] These concepts are constitutive of the good life, but they rarely appear in our conversations around global health. Given the Church's significant presence in low-income settings, it would do well to devote more energy to these concepts not simply because they are religious but because they would help solve genuine problems in global public health. Perhaps even more important, there is nothing overtly sectarian or exclusionary about more intentionally asking people to reflect on their gifts and where they can best be used. This contribution works well in a pluralistic environment.

Misallocation of Resources

Another question that endures in the field of global public health is why the resources are not better aligned with actual need. The problem is well-known: where the Global South experiences about 90 percent of the world's burden of disease, only about 10 percent of research resources are devoted to such issues.[15] There is also a well-worn history of failing to overcome the colonialist relationship between the Global North and Global South, where even good intentions cannot reorient the power relationship between the two.[16] These are not new observations, but little has been successful in placing those who are poor truly at the center of the field's work. This is not to suggest there is not a desire. There are heroic efforts, including by those who work in global public health, to shift resources toward geographical and epidemiological areas of greatest need. Much like the previous exploration of vocation, though, the challenge is not merely a technical shift of resources. Rather, the solution requires a compelling reason as to why such an imbalance is unacceptable in the first place.

Public health's primary ethical tool is social justice. Work by scholars such as Margaret Whitehead[17] and Jonathan Mann[18] clarified ideas that are central to the work of public health, such as equity and the right to health. Yet these ideas often lack a robust description of the human person. This might make arguments that are about caring

---

[14] See Shane Sinclair et al., "Compassion: A Scoping Review of the Healthcare Literature," *BMC Palliative Care* 15 (2016): doi.org/10.1186/s12904-016-0080-0.

[15] See Global Forum for Health Research, *The 10/90 Report on Health Research*, (Geneva: World Health Organization, 2000), announcementsfiles.cohred.org/gfhr_pub/assoc/s14791e/s14791e.pdf.

[16] See Jorge José Ferrer, SJ, "Research as a Restorative Practice: Catholic Social Teaching and the Ethics of Biomedical Research," in *Catholic Bioethics and Social Justice: The Praxis of US Health Care in a Globalized World*, ed. M. Therese Lysaught and Michael McCarthy (Collegeville, MN: Liturgical Press, 2019), 363–376; Lawrence O. Gostin, "Redressing the Unconscionable Health Gap: A Global Plan for Justice," *Harvard Law and Policy Review* 4 (2010): 271–294.

[17] See Margaret Whitehead, "The Concepts and Principles of Equity and Health," *International Journal of Health Services* 22, no. 3 (1992): 429–445.

[18] See Jonathan M. Mann, "Medicine and Public Health, Ethics and Human Rights," *Hastings Center Report* 27, no. 3 (1997): 6–13.

for human people in need less compelling to the broader public. Human rights can often become disembodied, focusing instead on rights-claiming or rights-respecting entities, or equity can sometimes become an impersonal calculation as to whether outcomes are distributed fairly. There is certainly a need for these ideas, but they have proven insufficient for shifting the allocation of global resources in the way that global public health clearly desires.

The moral tradition in the Catholic Church includes Catholic social teaching, or what it takes to build a just society.[19] A principal tenet of Catholic social teaching is a "preferential option for the poor." This idea suggests that, "God has a preferential option for the poor not because they are necessarily better than others, morally or religiously, but simply because they are poor and living in an inhuman situation that is contrary to God's will."[20] Prioritizing people who have the most barriers to achieving full human flourishing should be a global public health goal. This is the view that the Church puts forward but also has largely failed to convince the world to embrace. Hence, there is a dual benefit if the Church can partner with global health institutions to close the persistent 90/10 gap: it advances the positions of both entities. It would do so in a way that places the human person, especially those who are poor, at the center of concern.

Capacity Building

A final area where global public health might benefit from other disciplinary expertise is on the issue of capacity building. This topic focuses on strengthening local health systems so that they are more sustainable.[21] Efforts include educating a local health workforce, creating local infrastructure and supply chains, building the needed information technology, instituting appropriate financial systems, supporting local governance, and much more. At a simple level, this would force short-term medical missions to consider why their volunteers take medical histories and blood pressure rather than training locals to learn these very transferrable skills.[22] On a bigger

---

[19] See Thomas Massaro, SJ, *Living Justice: Catholic Social Teaching in Action*, 3rd ed. (Lanham, MD: Rowman & Littlefield, 2016).

[20] Gustavo Gutiérrez, *On Job: God-Talk and the Suffering of the Innocent*, trans. Matthew J. O'Connell (Maryknoll, NY: Orbis Books, 1987), 94.

[21] See World Health Organization, "Everbody's Business: Strengthening Health Systems to Improve Health Outcomes," www.who.int/healthsystems/strategy/everybodys_business.pdf; Tamara Hafner and Jeremy Shiffman, "The Emergence of Global Attention to Health Systems Strengthening," *Health Policy Plan* 28, no. 1 (2013): 41–50.

[22] See Stephanie D. Roche et al., "International Short-Term Medical Missions: A Systematic Review of Recommended Practices," *International Journal of Public Health* 62, no. 1 (2017): 31–42; Melissa K. Melby et al., "Beyond Medical 'Missions' to Impact-Driven Short-Term Experiences in Global Health (STEGHs): Ethical

scale, it asks why global philanthropy and government aid allocates resources to buildings that they can put names on but that will never be staffed instead of sewage systems and electrical grids that provide the needed foundation for further development.[23] The lack of motivation for capacity building has some of the same roots as our overinvestment in medical care and underinvestment in public health. It is hard to excite people about preventing illness in statistical lives rather than curing illness in an identifiable person.

At the same time, the Church has several resources in its tradition that should provide motivation to focus on building local capacity in the world of global public health. When considering the issues of capacity building, some might suggest subsidiarity as the cornerstone concept—that what can be done by local structures ought to be done by them. But an even more fundamental motivation for capacity building than subsidiarity is human dignity. On the one hand, when global institutions refuse to invest in local resources and local knowledge what they are essentially saying is that the local realities are not good enough. They cannot be trusted, or they are not sufficient. This is not just a poor long-term strategy for global health; but it tears at the dignity of those who live, work, play, and pray in those communities. On the other hand, when global institutions recognize resources that exist in communities of need and when they trust that making the health issues that characterize the Global North less dominant as possible, they are not only improving health indicators but also recognizing the dignity and the capacity that were always there.

GLOBAL PUBLIC HEALTH'S INSIGHTS FOR THE CHURCH

Some may suggest that the Catholic Church is already doing a great deal more with global public health than this essay implies. They could rightly point to the work done by the local churches in low-income countries or done by Catholic Relief Services in many of these same settings. Catholic Medical Mission Board, innumerable religious congregations, and the Catholic Health Association are just some of the many organizations whose work might lead one to believe that the Church is fully engaged in global public health. However, this position does not address the more fundamental issue presented by this essay. It is not a matter of whether the Church is doing enough but whether

---

Principles to Optimize Community Benefit and Learner Experience," *Academic Medicine* 91, no. 5 (2016): 633–638.

[23] See Bruno Marchal et al., "Global Health Actors Claim to Support Health System Strengthening: Is This Reality or Rhetoric?" *PLoS Medicine* 6, no. 4 (2009): e1000059; Ashley E. Warren et al., "Global Health Initiative Investments and Health Systems Strengthening: A Content Analysis of Global Fund Investments," *Global Health* 9, no. 1 (2013): doi.org/10.1186/1744-8603-9-30.

the Church is using its resources as wisely as possible. Perhaps this approach requires shifting material resources to new activities, but it may also mean applying intellectual and spiritual resources to major questions facing global public health.

Those in the Church must face the fact that most of its resources—time, money, and intellectual energy—that are devoted to health are focused on acute care of individuals who are ill. There are good reasons for heavy investment in this work,[24] but the Church may be doing so without considering the full range of possibilities for its limited resources. Throughout history, the Church has been central in caring for victims of epidemics, such as the plague in sixteenth century Europe[25] and cholera outbreaks in the nineteenth century United States.[26] But what if efforts could have prevented the plague in the first place? As Ebola or the novel coronavirus move more easily through the globalized world, we should not need much convincing that the Church should be asking the modern equivalent of that question wherever it is.

Greater engagement between global public health and the Catholic Church would also shape the Church itself. First, the collaboration would likely expand the Church's moral resources or at least require the Church to do some rebalancing of the moral issues that capture its time and attention. The corporal works of mercy, for example, are nearly always described as taking care of an individual's acute need. While this is good and holy work, perhaps the Church might also consider the ways in which the corporal works of mercy can be lived out at the population level. Mercy might not only require one to give drink to the thirsty but to work for policies that ensure potable water systems for an entire community. The works of mercy are a touchstone for the ministry of the Church, and their core ideas are widely embraced in the world of global health. Yet in widening one's moral imagination, even greater possibility for collaboration might emerge. This is true for many foundational moral teachings in Scripture and tradition.

As we read in the Gospel, "For I was hungry and you gave me food, I was thirsty and you gave me drink, a stranger and you welcomed me,

---

[24] See Halley S. Faust, and Paul T. Menzel, "Our Alleviation Bias: Why Do We Value Alleviating Harm More Than Preventing Harm?" in *Prevention Vs. Treatment: What's the Right Balance?*, ed. Halley S. Faust and Paul T. Menzel (Oxford: Oxford University Press, 2011), 139–175; Dan W. Brock, "Identified Versus Statistical Lives: Some Introductory Issues and Arguments," in *Identified Versus Statistical Lives*, ed. Nir Eyal, I. Glen Cohen, and Norman Daniels (Oxford: Oxford University Press, 2015), 43–52.

[25] See John O'Malley, *The First Jesuits* (Cambridge, MA: Harvard University Press, 1993).

[26] See Margaret M. McGuinness, *Called to Serve: A History of Nuns in America* (New York: New York University Press, 2013).

naked and you clothed me, ill and you cared for me, in prison and you visited me" (Matthew 25: 35–36). These actions rightly focus the moral concerns of the Christian faith, and they should not be diminished in any way. However, the Church might also consider the ways in which mercy can be fostered through organizations or policies. This expansion would be akin to the way public health is a partner to clinical medicine. Public health's role is not to replace medical care but to suggest that there are other ways of achieving health, and sometimes they are even more effective than treating the acute illness of individuals. In the same way, without diminishing corporal works of mercy that take place between two individuals, policies of mercy may sometimes be a preferable use of limited resources. In this way, the insights of public health help expand the Church's moral horizon. Hence, we could say: "For I was hungry, and you fought climate change and did not destroy our farms and fisheries. I was thirsty, and you built infrastructures to guarantee safe drinking water; a stranger and your laws allowed me to find asylum; naked and local industry produced my clothing; ill and you educated my local health worker; in prison and the system rehabilitated me."

The benefits of greater collaboration do not just include this expansion of the moral imagination of the Church, as important as that would be. This engagement may initiate a healthy degree of conversion within the Church on several matters. Some of the worst actors in short-term medical mission are faith-based organizations. Too often, these groups undermine local health systems or violate standard ethical norms,[27] all under the auspices of doing good. In addition, some of the most invincible power structures can be found within the Church itself. Therefore, as this essay suggests that power in global public health must shift to the Global South, one might rightly observe that the same shift in power and voice would benefit the global Church. Finally, the Church's narrow focus on a small set of moral issues often obscures the rich tapestry that exists in the Church's moral tradition.[28] The collaboration between global public health and religious organizations often gets stymied because of disagreements on reproductive health or gender issues. At the same time, there is vast agreement on matters rooted in Catholic social teaching, including human dignity, the common good, and the option for the poor. Such an approach does not ignore areas where there is

---

[27] See Virginia Rowthorn et al., "Not above the Law: A Legal and Ethical Analysis of Short-Term Global Health Experiences," *Annals of Global Health* 85, no. 1 (2019): doi.org/10.5334/aogh.2451.

[28] See James F. Keenan, *A History of Catholic Moral Theology in the Twentieth Century: From Confessing Sins to Liberating Consciences* (London: Continuum, 2010).

legitimate disagreement, but it suggests that we might promote a different starting point and allow trust to build over time.

CONCLUSION

The central premise of this essay is that global public health has some enduring challenges that are not just a matter of technical ability but of disciplinary identity. Above, I describe examples of areas that should be explored for global public health to achieve its fullest potential in the decades to come. Some may suggest that such challenges are not as significant as this essay would lead readers to believe. They might rightly point to the many legitimate successes of global public health over the past couple decades, such as increased vaccination rates[29] or decreased infant mortality rates.[30] At the same time, the field of global public health has not yet figured out how to successfully position itself alongside the medical professions, even though they often compete for the same limited resources.

Even if one accepts the premise of this essay—that global public health should directly address issues of identity—one might still resist that one of the solutions is to look to a religious organization such as the Catholic Church for solutions. Such skeptics would rightly ask if deeper engagement of the Catholic Church is truly beneficial for global health. They could observe that the Church does important work in improving health but might suggest that it is possible for the Church to continue to do its work of caring for the sick and allow other institutions, which may be better equipped, to do the work of public health. This is a fair concern, but given the resources and influence of faith-based organizations around the world,[31] this is just a theoretical suggestion. Those in public health may push for government or secular organizations in developing economies to take on larger roles, but the poorest countries in the world are a good distance from that being a plausible solution. It seems reasonable, then, that the Catholic Church and other faith-based organizations might be true thought partners in achieving global health objectives.

It is quite intentional that the three suggestions above are not technical in nature. Global public health would never rely on the theological insights of the Catholic Church for matters of epidemiology or supply chain management. Instead, each of the three speak to what it means to be human or what it takes to build a vibrant

---

[29] See Samantha Vanderslott et al., "Vaccination," (2019), ourworldindata.org/vaccination.
[30] See World Health Organization, "Infant Mortality: Situation and Trends," www.who.int/gho/child_health/mortality/neonatal_infant_text/en/.
[31] See Jeffrey Haynes, *Faith-Based Organizations at the United Nations* (New York: Palgrave Macmillan, 2014); José Casanova, *Public Religions in the Modern World* (Chicago: University of Chicago Press, 1994).

human community. Those are areas in which the Catholic Church and other religious communities have longstanding insights which can be informative to global public health and the broader world.

The first suggestion is that the human person, including the global public health professional, must cultivate a sense of vocation or purpose. Without a grounding in vocation, the often-challenging work of global public health can become burdensome and joyless. Second, if global public health is going to advocate for a global community, it must provide more compelling reasons why the resources of global health should be more justly distributed. Global health research must be more responsive to needs than it is to power. Such a shift does not come with political agreements alone but instead with a broadly communicated, compelling vision of the global human community. Third, the field must place the dignity and capacity of those in the Global South at the center of all it does. If genuine belief in human dignity is the starting point of all global public health, those living in developing economies will less often be objects needing assistance and instead be subjects with the capacity for leading change. At their core, all three of these depend on choices and lead to moral acts. These choices and actions are informed by how individuals choose to see the human person and shape the human community.

Several resources from within the Catholic Church would prove beneficial to enduring issues faced by the global public health community. At the same time, the engagement would almost certainly benefit the Church as well. Pope Francis, through appointments of cardinals, encyclicals and other writings, and his insistence on synodality, has pushed for greater power to reside in the Global South. He has also insisted on the capacity of those in developing economies, and he has made joy a hallmark of his pontificate. Nevertheless, the proposed shift will demand a long journey for the Church. Collaboration between global public health and the Catholic Church could, and should, change institutions in both realms. Nevertheless, those in global public health and the Catholic Church would both suggest that they are most interested in any positive effects being not for the institutions themselves but for those who are suffering unnecessarily around the world. This shared desire is yet another area of common ground that suggests a stronger relationship between global public health and the Catholic Church could bear much fruit.
M

Michael D. Rozier, SJ, is assistant professor of Health Management and Policy at the College for Public Health and Social Justice at Saint Louis University, with a secondary appointment in the Albert Gnaegi Center for Health Care Ethics. He earned his PhD in the Department of Health Management and Policy at the University of Michigan. After receiving his Master of Public Health degree at Johns Hopkins University, in 2008, Fr.

Rozier worked as an ethics fellow with the World Health Organization in Geneva, Switzerland. From 2008 to 2011, he was an instructor at Saint Louis University where he taught courses in global health, health and justice, and public health ethics. He was also the founding director of the College for Public Health and Social Justice's undergraduate degree in public health and oversaw service-learning activities, including several trips abroad with students. His peer-reviewed work has been published in the *American Journal of Public Health, Journal of Public Health Management and Practice, Journal of Religion and Health*, and *Public Health Ethics*, among others. He is also a regular contributor on matters of public health and health policy to *America* magazine and *Health Progress*, a publication of the Catholic Health Association. His areas of research focus on goal-setting and resource allocation in low-income countries, the relationship of medical missions to the local health systems they serve, and the ways public health ethics frames health challenges differently than medical ethics. Fr. Rozier lived and worked in Canada, Switzerland, and throughout Latin America.

The Affordable Care Act and Pharmaceuticals: An Economic Perspective

Tracy L. Regan

AS EVIDENCED BY THE NUMEROUS Democratic presidential debates, health care continues to be a top priority for the US, but it is also a source of much disagreement and differing opinions. A recent poll by the Kaiser Family Foundation (KFF) indicates that a majority of Democrats consider new legislation and oversight of our health care system to be of the upmost concern. Regardless of one's political affiliation, Americans indicate that the cost and affordability of health care need to be addressed.[1] Despite a campaign promise to "Replace and Repeal" Obamacare, President Trump failed to do so. Former Representative Paul Ryan was unsuccessful in bringing his Affordable Health Care Act (AHCA), in its many iterations, to a vote as it was universally criticized. Ryan was, however, able to pass the 2017 Tax Cuts & Jobs Act that, along with overhauling the tax code of the past three decades, made the Shared Responsibility Payment no longer applicable in 2019.[2]

The individual mandate has been the most contentious part of the 2010 Affordable Care Act (ACA, aka Obamacare) despite the Supreme Court upholding it as a "tax" in 2012 and deeming it to be constitutional.[3] This requirement was originally supported by the Heritage Foundation as a way to stress personal responsibility and to eliminate the "free rider" issue surrounding the uninsured. Baicker and Chandra note that

> Insurance, in its simplest form, works by pooling risks.... The premium for health insurance is the expected cost of treatment for everyone in the pool. The key insight is that not everyone will fall sick

---

[1] See Ashley Kirzinger et al., "KFF Health Tracking Poll—November 2018: Priorities for New Congress and the Future of the ACA and Medicaid Expansion," (2018), www.kff.org/health-reform/poll-finding/kff-health-tracking-poll-november-2018-priorities-congress-future-aca-medicaid-expansion/.
[2] Several states have continued to enforce the individual mandate, Massachusetts being one of them.
[3] Former President Obama initially did not want people referring to the ACA as "Obamacare" but soon warmed up to the nickname as he was incredibly proud of this domestic achievement.

at the same time, so it is possible to pay for the care of the sick even though it costs more than their premiums. This is also why it is particularly important for people to get insured when they are healthy—to protect against the risk of needing extra resources to devote to health care if they fall ill.[4]

Regardless of how one feels about the ACA, it did succeed in lowering the rate of uninsurance from its peak in 2013 of 18 percent to a low of 10.9 percent three years later when Trump was elected.[5] Despite its success, Obama will be the first to admit that the ACA is not perfect. Obama notes that "there is more work to do to ensure that all Americans have access to high quality, affordable health care. ... Rather than jeopardize financial security and access to care for tens of millions of Americans, policymakers should develop a plan to build on what works before they unravel what is in place."[6]

Having health insurance does not guarantee health[7] but the "reckless"[8] efforts the Trump administration has made to let Obamacare fail will most certainly undermine its success.[9] A recent Gallup poll reports that in the last quarter of 2018, the uninsured rate increased to 13.7 percent.[10] It certainly does not seem that we are "Making America Great Again" by returning to pre-ACA rates of uninsured.

To date, there are still fourteen states that have not expanded their Medicaid program. A quick glance at a map of the US will reveal the seemingly perpetual northern-southern divide in this country. The KFF estimates that 2.3 million Americans are in the "coverage gap," 92 percent of these people live in the South, and a majority reside in either Texas or Florida. There is a disproportionate amount of racial and ethnic minorities (i.e., Hispanic and Black) in these two states that

---

[4] See Katherine Baicker and Amitabh Chandra, "Myths and Misconceptions About U.S. Health Insurance," *Health Affairs* 27, no. 6 (2008): doi/10.1377/hlthaff.27.6.w533.
[5] See Dan Witters, "U.S. Uninsured Rate Rises to Four-Year High," news.gallup.com/poll/246134/uninsured-rate-rises-four-year-high.aspx.
[6] Barack H. Obama, "Repealing the ACA without a Replacement: The Risks to American Health Care," *New England Journal of Medicine* 376, no. 4 (2017): 297–299, at 297.
[7] For example, see Baicker, and Chandra, "Myths and Misconceptions."
[8] Obama, "Repealing the ACA without a Replacement," 297.
[9] Other efforts by the Trump administration include: 1) funding cuts for enrollment advertising; 2) a shorter sign-up window for individuals to secure health insurance; 3) the shutting down of healthcare.gov for 12-hour blocks on Sundays to do "maintenance"; 5) the terminating of contracts for health care "in person assisters" in 18 cities; and 6) the halting of participation in enrollment events, especially those in the South and those that would have attracted Hispanics and Blacks.
[10] See Witters, "U.S. Uninsured Rate Rises to Four-Year High."

are run by Republican governors.[11] The lack of bipartisan effort in Congress is clearly hurting this country's citizens.

While the ACA is often referred to as "health care reform," I would argue that there was nothing particularly revolutionary about it.[12] It preserved our employer-based system of health insurance in the US.[13] Nearly half of Americans still get their health insurance from their employer.[14] However, hundreds of millions of Americans are dependent on the government for their health insurance. The Centers for Medicare and Medicaid Services (CMS) report that seventy-one million individuals were enrolled in Medicaid and Chip in 2019 and sixty million senior citizens received their coverage through Medicare.[15] A Gallup poll from 2018 indicates that 79 percent of Medicaid/Medicare patients rate their healthcare coverage as "excellent" or "good." The popularity and success of Medicare may provide insight into why Democratic Presidential candidates Senator Bernie Sanders and Senator Elizabeth Warren were campaigning on platforms that would provide "Medicare for All," while Mayor Pete Buttigieg's slogan was "Medicare for All Who Want It." Former Vice President, front-runner, and now President Joe Biden is supporting the

---

[11] See Rabah Kamal et al., "What Are the Recent and Forecasted Trends in Prescription Drug Spending?" www.healthsystemtracker.org/chart-collection/recent-forecasted-trends-prescription-drug-spending/#item-start.

[12] Some of the popular features that addressed the former abuses by the health insurance industry include the: 1) end to pre-existing condition discrimination; 2) end to limits on care; 3) dependent coverage until age 26; and 4) end to coverage cancellations. It is worth noting that the first three items mentioned were part of the Vote-O-Rama where Republicans attempted to overhaul these key features of the ACA on the night of Trump's inauguration.

[13] This is likely because Obama learned from the Clintons' efforts to create universal health care coverage in a very complicated manner, which were ultimately defeated by the Right, who ran advertisements against them. These ads scared Americans that they would lose their employer-sponsored health care. Moreover, the traditional pairing of wage/salary employment and health benefits is an accident of history. In order to compete for domestic workers during World War II, many employers got around the wage controls that were employed by offering fringe benefits. The tax benefits associated with contributions to an employee's health insurance plan—both by the employer and employee—became codified into law and have cost the federal government billions of dollars in missed revenue, while saving companies money and ensuring that they lobby to continue these protections. See Melissa A. Thomasson, "From Sickness to Health: The Twentieth-Century Development of U.S. Health Insurance," *Explorations in Economic History* 39, no. 3 (2002): 233–253.

[14] See Kaiser Family Foundation, "Health Insurance Coverage of the Total Population," www.kff.org/other/state-indicator/total-population/?current Timeframe=0&sortModel=%7B%22colId%22:%22Location%22,%22sort%22:%22 asc%22%7D.

[15] See Centers for Medicare & Medicaid Services, "National Health Expenditure Data Fact Sheet," www.cms.gov/Research-Statistics-Data-and-Systems/Statistics-Trends-and-Reports/NationalHealthExpendData/NHE-Fact-Sheet.

introduction of a public option into the ACA.[16] The Republicans have resumed their "socialized medicine" fearmongering chants of the Obama years. Trump continued to stoke such fears when he promised that we "will never let socialism destroy American healthcare" during his 2020 State of the Union. Regardless of President Biden's election to office in 2020, it is unlikely that a complete overhaul of our health care system will occur because of the divisiveness of this issue and our divided Congress.

AN INTERNATIONAL COMPARISON

The United Kingdom's National Health Service (NHS) is probably the truest version of "socialized" medicine; the NHS is both the payer and provider of health care. The NHS Constitution guarantees "the right to access care without discrimination" and operates with the core principle that care be "free at the point of use."[17] This ethos is echoed in France where the "provision of health care is a national responsibility"; their coverage is universal and compulsory.[18] Health insurance in France is provided at the employer level. This is similar to what Warren was proposing. Instead of having employers pay a private health insurance firm, they would instead pay the government. A single-payer system would likely reduce overall medical costs and provide savings, as it would lower administrative costs and help to reduce pharmaceutical prices.[19] The virtues of a nationalized health care system are clear, as the World Health Organization (WHO) ranked France as number one in its 2000 World Health Report of the health systems of its 191 member countries.[20] By comparison, the US ranked number thirty-four. This is despite the extraordinary amount of money we spend on health care.

The US spends the most, both in total and on a per capita basis, on health care relative to all OECD countries.[21] In 2018, the US spent $10,586 per capita on health; the OECD average was $4,070.[22]

---

[16] Obama urged policy makers to introduce a public option in areas lacking individual market competition. See Obama, "Repealing the ACA without a Replacement."
[17] The Commonwealth Fund, "International Health Care System Profiles," www.commonwealthfund.org/international-health-policy-center/system-profiles.
[18] The Commonwealth Fund, "International Health Care System Profiles."
[19] A recent article in *Time* reported that in 2017 the US spends $2,500 per capita on health care administrative costs. This is about 34 percent of total health care expenditures. See Abigail Abrams, "The U.S. Spends $2,500 Per Person on Health Care Administrative Costs. Canada Spends $550. Here's Why," *Time*, January 6, 2020, www.time.com/5759972/health-care-administrative-costs/.
[20] By comparison, the U.K. ranked fourteenth.
[21] The OECD is the intergovernmental Organisation for Economic Cooperation and Development, which gathers 37 member countries to stimulate economic progress and world trade.
[22] Of the OECD countries, Switzerland ranked second highest in its per capita spending, $7,317, and Mexico was last at $1,138. See Organisation for Economic

America spent a total of $3.6 trillion in national health expenditures and this accounted for 17.7 percent of our Gross Domestic Product (GDP)![23] Unfortunately, this incredible level of spending does not translate into good health outcomes. The life expectancy at birth for the US is 78.6 years. This is twenty-eighth amongst OECD countries. We are just ahead of Estonia (78.2) and just behind the Czech Republic (79.1). By comparison, Japan spent $4,766 per capita in 2018 and has the highest life expectancy of 84.2 years.[24] Similarly, the US ranks worst in terms of self-reported measures of obesity or overweight, with 65.5 percent of the population ages fifteen and above.[25]

The American health care system—often referred to as a "disease-maintenance system"—fails in basic dimensions like preventative care and reimbursement schemes for physicians, but we do quite well when it comes to serious diseases like cancer. The innovation, technology, and research of the US has enabled many people to live longer and healthier lives. Many people from all over the world seek out care in the best hospitals that the US has to offer. The decentralized, for-profit, complicated system of health care has enabled much of this to happen along with the focus on innovation and firm creation in a capitalistic economy. The pharmaceutical industry plays a large part in all this as it attracts the admiration and ire of patients and politicians.

THE PHARMACEUTICAL INDUSTRY

In the US, 12 percent of our health spending goes to pharmaceuticals.[26] Our spending on pharmaceuticals easily outpaces the rate of inflation.[27] Many Americans are shielded from the full retail price of a pharmaceutical due to the cost-sharing features associated

---

Cooperation and Development, "Health Spending," data.oecd.org/healthres/health-spending.htm.
[23] See Centers for Medicare & Medicaid Services, "National Health Expenditure Data Fact Sheet."
[24] There are most certainly differences between these countries that help to explain some, but not all, of the discrepancies. See Organisation for Economic Cooperation and Development, "Life Expectancy at Birth," data.oecd.org/healthstat/life-expectancy-at-birth.htm.
[25] Alternatively, when measured rates of obesity are used instead, the rate is 71 percent. Using this metric, Chile ranks highest at 74.2 percent but has recently undergone a sweeping campaign against obesity that involves warning labels on packaged food, indicating high calorie and sugar content. See, for example, Andrew Jacobs, "In Sweeping War on Obesity, Chile Slays Tony the Tiger," *The New York Times*, February 7, 2018, www.nytimes.com/2018/02/07/health/obesity-chile-sugar-regulations.html.
[26] See Organisation for Economic Cooperation and Development, "Health Spending."
[27] For example, see Tracy L. Regan, "Generic Entry, Price Competition, and Market Segmentation in the Prescription Drug Market," *International Journal of Industrial Organization* 26, no. 4 (2008): 930–948.

with many prescription drug plans.[28] Despite this, Peterson-KFF Health System Tracker reports that nearly 25 percent of adult Americans have a difficult time affording their medicine.[29] It is not uncommon for cancer drugs to cost over $10,000 per month.[30] The high retail prices are somewhat the product of the system of patent protection that we provide in the US. A branded firm is able to secure a twenty-year patent on its drug, and during this period, it can employ monopoly pricing to reap back its multi-billion dollar risky investment in research, development, testing, and clinical trials that are all needed before a drug goes to market.[31] DiMasi et al. also estimate that the cost of bringing a pharmaceutical to market is $2.870 billion (2013 USD) with an "effective patent life" of only 10.6 years, due to the Food and Drug Administration's (FDA) stringent testing.[32] Patent protection is provided to ensure that pharmaceutical firms continue to innovate, but the therapeutic benefit associated with some molecular manipulations and strategic actions by the industry just result in higher prices. This situation couples with the recent emergence of Pharmaceutical Benefit Managers (PBMs) that have been flying below the radar as they disrupt the traditional supply chain. Their involvement in every juncture from manufacture to distribution often inflates the cost, and any price discounts they have secured often do not get passed down to the patient.

The US along with New Zealand are the only two countries that allow direct-to-consumer-advertising (DTCA). In 1997, spending on pharmaceutical marketing was $17.7 billion while in 2016 it exploded to $29.9 billion; half of the spending in 2016 was due to DTCA.[33] While the FDA tests for safety and efficacy, the FDA does not consider cost effectiveness and therapeutic benefit.[34] Recently there have been some outrageous cases of pharmaceutical pricing. Some of the recent include Gilead's cure for Hepatitis C, Sovaldi, which in 2015, when it was introduced to the market, cost $84,000 for a twelve-

---

[28] Co-insurance brings better awareness to the price of a pharmaceutical than co-payments do. Moreover, it is common knowledge that the list price of pharmaceuticals often does not reflect the true "price," due to the various negotiations and agreements between the manufacturers, the Pharmaceutical Benefit Managers (PBMs), and the insurers.
[29] See Kamal et al., "What Are the Recent and Forecasted Trends?"
[30] See Robert Langreth, "Drug Prices," www.bloomberg.com/quicktake/drug-prices.
[31] It should be noted that there are multiple patents filed on a pharmaceutical's behalf.
[32] The 1984 Waxman-Hatch Act was passed to address this, but some of its provisions have been abused and manipulated by pharmaceutical firms. See Joseph A. DiMasi et al., "Innovation in the Pharmaceutical Industry: New Estimates of R&D Costs," *Journal of Health Economics* 47 (2016): 20–33.
[33] See Lisa M. Schwartz, and Steven Woloshin, "Medical Marketing in the United States, 1997-2016," *JAMA* 321, no. 1 (2019): 80–96.
[34] The National Institute for Health and Clinical Excellence (NICE) in England sets guidelines for clinically effective treatments.

week regimen.³⁵ The life-saving treatments of insulin and EpiPens have made headlines of late as well. There is nothing restraining or regulating the prices of pharmaceuticals in the US, and only sometimes does public outcry lead to change.³⁶

There are various pricing strategies that are employed by countries around the world. These include organizations like NICE in the U.K. that establish the cost-effectiveness or value pricing of a drug. Germany employs reference pricing, whereby the insurer establishes a maximum payment that it will contribute, and the difference is the consumer's responsibility. While problematic, developing countries like El Salvador instituted price controls.³⁷ Due to the universal nature of most (developed) countries' health care systems, the government acts as the monopsony buyer and has full control over what drugs enter a market. Governments can use this power as leverage to negotiate much lower prices.³⁸ Regulating pharmaceutical prices in the US is not straightforward, as we produce nearly 60 percent of the world's drugs.³⁹

CONCLUSION

There are models throughout the world that the US could look to in reforming its health care system and pharmaceutical industry. While politicians are ultimately the ones who pass laws, one must wonder if progress is achievable because issues of health insurance have become so politicized. We are living in a time when basic facts are being questioned and news now come in the form of tweets in lieu of credible journalism and data-driven evidence. While there seemed to be consensus that having nearly 50 million Americans without health insurance was concerning in the pre-ACA area, our country still has not agreed whether health care is a right of citizenship or not. While we continue to commit nearly 30 percent of our taxes to fund programs

---

[35] See Editorial Board, "Costly Hepatitis C Drugs for Everyone?" *The New York Times*, September 2, 2015, www.nytimes.com/2015/09/02/opinion/costly-hepatitis-c-drugs-for-everyone.html.

[36] There is another conversation to be had over the Sackler family's involvement in the opioid epidemic ravaging this country.

[37] Margaret Kyle speaks to the lack of entry into markets in the E.U. with price controls. See Margaret K. Kyle, "Pharmaceutical Price Controls and Entry Strategies," *Review of Economics and Statistics* 89, no. 1 (2007): 88–99.

[38] Comparing prices of pharmaceuticals in the US to those of other countries is not straightforward as the set of drugs can differ along with the institutional features of the health care system and the tax structure. This also couples with a difference in expectation and philosophy, as Americans often want the newest and "best" drug that is available.

[39] It is often thought that the US subsidizes the rest of the world's consumption of pharmaceuticals and the research and development (R&D) required to bring the drugs to market. See Ross C. DeVol et al., *The Global Biomedical Industry: Preserving U.S. Leadership* (Santa Monica, CA: Milken Institute, 2011).

like Medicare and Medicaid, we worry about the overreaching arc of the federal government.[40] We think the private sector is more efficient than the public sector. However, in the case of healthcare, I would argue that the private nature of the industry has left too many people behind and allowed companies to profit off the poor health of this country. Maybe it is time for the "government to get its hands *on* our Medicare."[41] M

Tracy L. Regan is an applied microeconomist who earned her PhD from the University of Arizona in 2003. She joined the Economics Department at the University of Miami and was a Visiting Scholar and then full-time lecturer in the Department of Economics in the Eller College of Business at the University of Arizona. In 2013, Dr. Regan joined the Economics Department at Boston College as an Associate Professor of the Practice. She has extensive teaching experience with specializations in introductory microeconomics, labor economics, health economics, and industrial organization. Her research agenda combines her interests in labor economics—(gender) wage gaps, language acquisition, and the economics of the family—and on health economics—the pharmaceutical industry, the health insurance market, risky health behaviors, student identity, and adolescent academic outcomes. While publishing in prestigious journals, at Boston College Regan has become involved in a wide variety of activities and committees that include the Women's Center, the Athletics Department, the Lonergan Institute, the Center for Teaching Excellence, and the Gabelli Presidential Scholars Program.

---

[40] See National Priorities Project, "Where Your Tax Dollar Was Spent in 2018," www.nationalpriorities.org/analysis/2019/tax-day-2019/where-your-tax-dollar-was-spent-2018/.
[41] Bob Cesca, "Keep Your Goddamn Government Hands Off My Medicare!" *Huffington Post*, September 5, 2009, www.huffpost.com/entry/get-your-goddamn-governme_b_252326.

Journal of Moral Theology

Part 3:
Global Public Health Ethics

Social Structures and Global Public Health Ethics

Daniel J. Daly

AT ITS INCEPTION, THE FIELD OF PUBLIC HEALTH addressed the structures that causally contributed to illness, suffering, and premature death. The Public Health Act of 1848 was a landmark event in the genesis of the field, as it introduced structural regulations into the management of water and sewer systems of industrial-era England. A structural problem, the spread of infectious disease, warranted an effective structural solution, legislation that transformed the waste and water systems of the nation.[1] More recently, Paul Farmer has emphasized that the miserable health outcomes suffered by the poor are not accidents of history but rather, result from "structural violence." Structural violence pertains to the "historically given (and often economically driven) processes and forces that conspire—whether through routine, ritual or, as is more commonly the case, the hard surfaces of life—to constrain agency."[2] Public health must, then, be concerned with the social structures that harm the health of so many of the world's poor.

While many in the world still clamor for access to clean water and the sanitary disposal of waste, the twenty-first century presents two new global public health crises. The first is global warming, which, according to the Intergovernmental Panel on Climate Change, threatens food and water shortages for the poor, and it will cause mass migrations and large-scale animal extinctions.[3] The second is the lack of health workers in the Global South. In 2010, the World Health Organization (WHO) urged the nations of the Global North to stop recruiting medical professionals from the Global South. There the WHO identified the "brain drain" of trained doctors and nurses as a

---

[1] See Elizabeth Fee and Theodore M. Brown, "The Public Health Act of 1848," *Bulletin of the World Health Organization* 83, no. 11 (2005): 866–867.
[2] Paul Farmer, *Pathologies of Power: Health, Human Rights, and the New War on the Poor* (Berkeley: University of California Press, 2005), 40.
[3] See Intergovernmental Panel on Climate Change, "Global Warming of 1.5°C: Summary for Policty Makers," www.ipcc.ch/sr15/chapter/spm/.

significant cause of health worker shortages, which in turn further limited access to medical care in source nations.[4]

This chapter develops in three parts. Part one presents critical realist social theory. Critical realism provides an account of social reality which enables global public health ethicists to understand the causal mechanisms that perpetuate the suffering of the poor. Part two presents the ethical lens of the essay—the structures of virtue and vice. The structures of virtue and vice offer concepts with which to ethically categorize social structures. Part three presents and then analyzes the public health crises of global warming and health worker shortages in the Global South through the lens of the structures of virtue and vice. It then offers recommendations to transform vicious structures into virtuous ones.

THE CRITICAL REALIST ACCOUNT OF SOCIAL STRUCTURE

Because global public health ethics is concerned with how social life affects the health of populations, the discipline requires a coherent account of social structure. While rival accounts of social structure abound in social theory, I contend that critical realist social theory provides the best account.[5] Here I summarize the critical realist account of social structure in six propositions. First, structures are real. They exist independently of the human mind.[6] Further, structures are not the aggregation of all of the human actions in the world. While human actions can causally contribute to social structures, structures themselves exist independently of human actions.

Second, structures are "social relations among pre-existing social positions."[7] A structure is composed of the relations among

---

[4] See World Health Organization, "Global Code of Practice on the International Recruitment of Health Personnel," www.who.int/hrh/migration/code/WHO_global_code_of_practice_EN.pdf.

[5] For a defense of critical realism see David Cloutier, "What Can Social Science Teach Catholic Social Thought About the Common Good," in *Empirical Foundations of the Common Good: What Theology Can Learn from Social Science*, ed. Daniel K. Finn (New York: Oxford University Press, 2017), 170–207; Daniel J. Daly, *The Structures of Virtue and Vice* (Washington, DC: Georgetown University Press, 2021), chapter 3; Daniel K. Finn, *Consumer Ethics in a Global Economy: How Buying Here Causes Injustice There* (Washington, DC: Georgetown University Press, 2019); Daniel K. Finn, ed., *Moral Agency within Social Structures and Culture: A Primer on Critical Realism for Christian Ethics* (Washington, DC: Georgetown University Press, 2020); Matthew Allen Shadle, *Interrupting Capitalism: Catholic Social Thought and the Economy* (New York: Oxford University Press, 2018).

[6] John Searle defined realism as "the view that the world exists independently of our representation of it." See John R. Searle, *The Construction of Social Reality* (New York: Free Press, 1995), 153.

[7] Douglas Porpora, "Who Is Responsible? Critical Realism, Market Harms, and Collective Responsibility," in *Distant Markets, Distant Harms: Economic Complicity and Christian Ethics*, ed. Daniel K. Finn (New York: Oxford University Press, 2014), 3–24, at 14; Margaret Archer, "Structural Conditioning and Personal Reflexivity," in

differentiated social positions, such as patient and physician. For example, the social position of a cardiac surgeon pre-dates a person's assumption of that position. Once a surgeon, the person enters into an already-established relation to the social position of the patient.

Third, a single structure is composed of a web or a network of relations.[8] The health care system in the United States is composed of patients, physicians, nurses, hospital administrators, medical insurance administrators, and many others. The patient-physician relation does not, by itself, compose the health care system. There are other relations, such as among administrators and physicians, physicians and medical insurance underwriters, and patients and medical insurance auditors, which interrelate to compose the structure. In short, structures are complex.

Fourth, structures causally influence the moral actions that individual agents freely choose. Each social position contains prescribed and proscribed practices and activities. Some actions will be rewarded and enabled, while others will be punished and constrained. Surgeons are enabled and rewarded for performing surgery and are constrained and often punished for routinely recommending non-surgical treatment.[9] While a surgeon could choose to take a conservative approach to surgery, she should not expect to advance in her career if she does. Structures enable and constrain the agency of position-holders.

Fifth, structures causally contribute to outcomes in the world. Consider the fact that people often cannot assume the social position of "patient" if they are poor. "Patient" is reserved for those who have the financial resources to pay for treatment. As a result, the poor lack access to medical care and live less-healthy and shorter lives than the non-poor.[10] Structures causally contribute to human well-being or human suffering.

Finally, organizations are a type of structure.[11] An organization is a hierarchical social institution which contains clearly articulated authority positions and norms. Massachusetts General Hospital is an organization with a president, a hierarchy of management positions, and norms that guide the actions of those who work at the institution.

Distant Markets, Distant Harms: Economic Complicity and Christian Ethics, ed. Daniel K. Finn (New York: Oxford University Press, 2014), 25–53.

[8] See Pierpaolo Donati, "The Morality of Action: Reflexivity, and the Relational Subject," in *Distant Markets, Distant Harms: Economic Complicity and Christian Ethics*, ed. Daniel K. Finn (New York: Oxford University Press, 2014), 54–88, at 57.

[9] See Philip F. Stahel et al., "Why Do Surgeons Continue to Perform Unnecessary Surgery?," *Patient Safety in Surgery* 11(2017): doi.org/10.1186/s13037-016-0117-6.

[10] See Inaki Permanyer, and Nathalie Scholl, "Global Trends in Lifespan Inequality: 1950-2015," *PLoS One* 14, no. 5 (2019): doi.org/10.1371/journal.pone.0215742.

[11] See Dave Elder-Vass, *The Causal Power of Social Structures: Emergence, Structure and Agency* (Cambridge, UK: Cambridge University Press, 2010), 195.

STRUCTURES OF VIRTUE AND VICE

The structures of virtue and vice provide an appropriate ethical lens with which to analyze and judge the structures that promote or thwart the health of populations.[12] The structures of virtue and vice maintain that structures causally influence the moral character that individuals acquire and consistently produce social outcomes that promote or undercut human well-being. This chapter focuses on the latter aspect of the definition. For the remainder of the chapter, I refer to virtuous structures as "those webs of relations that consistently promote the normative dignity and well-being of all affected by those relations, especially the dignity and well-being of the vulnerable. These are structures that encourage decisions that contribute to social justice, and its fruit, the common good."[13] Vicious structures "are webs of relations that consistently violate normative dignity, human well-being, especially that of the vulnerable. Also, vicious structures promote social injustice and undermine the common good."[14]

This definition requires two clarifications. Inherent dignity pertains to the transcendent value that each person has, while normative dignity morally requires agents to *actively respect* all persons as transcendently, not instrumentally, valuable.[15] Further, normative dignity is systematically respected in situations in which the common good has emerged. The Second Vatican Council's pastoral constitution, *Gaudium et Spes*, defines the common good as "the sum of those conditions of social life which allow social groups and their individual members relatively thorough and ready access to their own fulfillment" (no. 26). The common good exists if and only if all persons have "thorough and ready access" to the goods needed to live a properly human life: food, shelter, clothing, primary medical care, education, the ability to participate in the political life, and freedom of speech/thought/religion. In the contemporary world, this kind of easy access to needed goods results from social structures that enable such access. Today, the common good is a global reality. The social structures that promote or undermine the common good exist on a global scale and influence every nation on earth.

GLOBAL PUBLIC HEALTH CRISES

Global warming and the dearth of health workers in the Global South are global public health crises substantially caused by social structures. In this section, I describe how social structures causally

---

[12] Michael Rozier has argued likewise in Michael D. Rozier, "Structures of Virtue as a Framework for Public Health Ethics," *Public Health Ethics* 9, no. 1 (2016): 37–45.
[13] Daly, *The Structures of Virtue and Vice*, chapter 6.
[14] Daly, *The Structures of Virtue and Vice*, chapter 6.
[15] See Darlene Fozard Weaver, "Christian Anthropology and Health Care," *Health Care Ethics USA* 26, no. 4 (2018): 1–6.

contribute to each crisis and then suggest ways in which these vicious structures can be transformed into virtuous ones.

The Carbon Economy and Global Warming

Because other authors in this volume document the public health harm caused by global warming,[16] this chapter prescinds from such a discussion and instead considers how global warming is structurally-caused. "The eco-conscious CEO" provides an example of how structures contribute to global warming. Imagine that a CEO, Kate, wanted to transform her energy-producing corporation by moving from the use of fossil fuels to the production of renewable energy. While she concedes that this transition could decrease profits and market value for the foreseeable future, Kate has chosen this path because she has recently become aware that global warming is a public health crisis that will disproportionately harm the world's most impoverished peoples.[17]

However, once Kate assumed the position of CEO in this organization, she entered relationships with: a board of directors, who can terminate her; shareholders, who can lobby for her to be terminated; and, governmental regulators, who can enforce corporate laws. One law that the board, the shareholders, and regulators will be keen to enforce is the "maximize rule." James Gamble, a former corporate attorney, discussed this rule in a 2019 article entitled, "The Most Important Problem in the World." Gamble wrote that the maximize rule is the most critical problem in the world because corporate management, including CEOs and boards of directors, is "obligated by law to make decisions that maximize the economic value of the company."[18] Other goods, such as the well-being of the poor or the environment, cannot legally justify a corporate decision. Thus, these goods are ignored and often subsequently harmed.

The maximize rule effectively constrains Kate's actions and practices as CEO. As CEO, Kate must demonstrate that her decisions maximize the economic value of her organization. If her actions do not maximize value but instead help to heal the planet, the board will fire her, and shareholders will sue her. While Kate is free to attempt

---

[16] In this book, see the chapters of Walter Ricciardi and Laura Mancini, Keith Martin, and Philip Landrigan.

[17] See Intergovernmental Panel on Climate Change, "Climate Change 2014: Impacts, Adaption, and Vulnerability," www.ipcc.ch/site/assets/uploads/2018/02/ar5_wgII_spm_en.pdf. The report notes that "People who are socially, economically, culturally, politically, institutionally or otherwise marginalised are especially vulnerable to climate change," 6.

[18] James Gamble, "The Most Important Problem in the World," www.medium.com/@jgg4553542/the-most-important-problem-in-the-world-ad22ade0ccfe.

restructuring the corporation, she will quickly meet punishments that will prevent her from doing so.

Here we see that the position of CEO is defined before Kate enters it. It is a pre-existing social position with prescribed practices and norms that Kate is expected to follow lest she suffers penalties, such as termination. In the end, due to laws and the pre-existing relations among the CEO, the board of directors, and shareholders, Kate is constrained from transforming the means by which the corporation produces energy.

Corporate governance is a vicious structure because it enables corporate position-holders to systematically ignore and exacerbate the suffering of the poor as well as contribute to the on-going destruction of the global ecosystem. However, structures can be transformed. Virtuous corporate structures would enable CEOs like Kate to "go green" and thereby promote the well-being of the poor and creation as a whole. Virtuous structures constrain harmful practices and enable practices that benefit the poor and vulnerable.

The structures that causally contribute to the warming of the globe will not be transformed only through the goodwill of individual people. Structural change, which emerges from the coordinated actions of many people, requires the redesigning of relations among social positions, such as board members and CEOs. Specifically, the maximize rule should be replaced with a rule that allows corporate management to account for human and ecological well-being in their decision making.[19] Only then, CEOs such as Kate would be enabled to enact their goodwill for the planet and the poor.

Global warming is a structural public health crisis that requires a structural moral solution. The task for Christians and all people of goodwill is to join in solidarity to transform the structures that warm the globe. According to St. John Paul II, individual agents practice the virtue of solidarity, which is the firm determination to commit oneself to the common good through "interdependence," "togetherness," and "collaboration." (*Sollicitudo Rei Socialis*, no. 39). According to Pope Francis, only then does solidarity "open the way to other structural transformations" (*Evangelii Gaudium*, no. 189).

Recruiting Health Workers in the Global South

The second structurally-caused public health crisis is the shortage of medical professionals in the Global South. Over eighty nations do not meet the World Health Organization's (WHO) minimum level of health care worker staffing. The situation is expected to worsen, as the WHO predicts that by 2035 there will be a global shortage of 12.3

---

[19] See Gamble, "The Most Important Problem in the World."

million health workers.[20] This is a significant problem because "the most critical issue facing health care systems is a shortage of people who make them work."[21] While South-East Asia has the largest numerical deficit of workers, Sub-Saharan Africa faces the most acute shortages. The WHO estimates that while Sub-Saharan Africa has twenty percent of the global disease burden, it has only three percent of the world's health workers.[22] More pointedly, Ghana has 1.8 physicians per 10,000 residents, Ethiopia 1, and Chad .475, while Cuba has 81, Switzerland 40, the United States 26.[23] Nurse ratios follow a similar pattern. These statistics attest to the uneven global distribution of health workers.

This inequality is not an accident of history but is the product of human values, decisions, and social structures. Like global warming, multiple social structures have contributed to this global public health crisis. Sub-Saharan Africa, for instance, currently lacks the institutions needed to train health care workers. In particular, "In the 47 countries of sub-Saharan Africa, just 168 medical schools exist. Of those countries, 11 have no medical schools, and 24 countries have only one medical school."[24] Yet another cause of this crisis is the active recruitment of health workers from developing countries to developed ones.

Physicians and nurses routinely train in developing nations, at taxpayer expense, only to be recruited by organizations in the United States, Canada, and other Global North nations. Medical institutions in the United States, for example, recruit internationally to address health worker shortages due to the aging of the "Baby Boomer" generation, population growth, and an increase in insured patients due to implementation of the Affordable Care Act. A 2020 study found that "physician shortages ... will likely increase over the next 10 years and may influence the delivery of healthcare, negatively affecting patient outcomes."[25] To address the impending crisis the authors of the study suggest "attracting foreign-trained doctors."

---

[20] See World Health Organization, "Global Health Workforce Shortage to Reach 12.9 Million in Coming Decades," www.who.int/mediacentre/news/releases/2013/health-workforce-shortage/en/.
[21] World Health Organization, "World Health Report: 2003," www.who.int/whr/2003/en/whr03_en.pdf?ua=1.
[22] World Health Organization, "World Health Report: 2003," 8.
[23] See World Health Organization, "Density of Physician Database," www.who.int/gho/health_workforce/physicians_density/en/.
[24] World Health Organization, "Global Health Workforce Shortage to Reach 12.9 Million in Coming Decades."
[25] Xiaoming Zhang et al., "Physician Workforce in the United States of America: Forecasting Nationwide Shortages," *Human Resources for Health* 18, no. 1 (2020): doi.org/10.1186/s12960-020-0448-3.

Social structures facilitate the pull of workers to the North. In 2016, the H-1B Specialty Occupations Visa Program permitted US employers to fill 10,500 physician positions.[26] In this program, health care positions are considered "specialty occupations." Crucially, some of the workers who are targeted hail from nations with severe health worker shortages. This legal structure is accompanied by a large and effective industry devoted to the active recruitment of health workers in the Global South. Over 160,000 foreign-born nurses who were actively recruited currently work in the United States.[27] In 2007, the Philippines had seventy-seven US-based nurse recruiting agencies, and India had fifty-six. For example, O'Grady Peyton International, the largest US-based international recruiter of health workers, has offices in New Delhi and Manila to recruit physicians and nurses.[28] As the United States continues to recruit Filipino nurses aggressively, the US enjoys a ratio of 8.6 nurses per 1,000 people, while the Philippines suffers from .2 nurses per 1,000.[29]

The "brain drain" of medical professionals from the Global South is staggering. Approximately twenty percent of all physicians and ten percent of nurses in the UK, US, Canada, and Australia received their medical training elsewhere.[30] Moreover, over eighty percent of trained medical professionals emigrate from Haiti, and nearly twenty-five percent of doctors and nurses trained on the continent of Africa currently practice in developed countries.[31] This social structure intentionally siphons workers toward the Global North and away from source nations, such as Ghana, India, and the Philippines.

Recruiting from nations with health workers shortages contributes to and exacerbates these shortages and lowers the standard of medical care provided in these nations. In 2010, the WHO wrote that it is "deeply concerned that the severe shortage of health personnel, including highly educated and trained health personnel, in many Member States, constitutes a *major threat to the performance of health systems* and undermines the ability of these countries to achieve the Millennium Development Goals and other internationally agreed

---

[26] See American Medical Association, "Issue Update," www.assets.ama-assn.org/sub/advocacy-update/2019-01-10.html.
[27] See Lawrence Gostin and Paula O'Brien, "Health Worker Shortages and Global Justice," www.milbank.org/publications/health-worker-shortages-and-global-justice/.
[28] See Edward J. Mills et al., "Should Active Recruitment of Health Workers from Sub-Saharan Africa Be Viewed as a Crime?" *Lancet* 371, no. 9613 (2008): 685–688. See also O'Grady Peyton International, "Employer FAQs," www.ogradypeyton.com/facilities/client-faqs.aspx.
[29] See World Bank, "Nurses and Midwives Per 1,000 People," www.data.worldbank.org/indicator/SH.MED.NUMW.P3.
[30] See Carwyn R. Hooper, "Adding Insult to Injury: The Healthcare Brain Drain," *Journal of Medical Ethics* 34, no. 9 (2008): 684–687.
[31] See Hooper, "Adding Insult to Injury."

development goals."[32] Unsurprisingly, the lack of health care workers harms public health.

Here we find a social structure—a web of relations among social positions—that enables hospitals in the United States to recruit and employ doctors and nurses from countries desperate to retain their few health workers, such as Ghana. Laws, such as the H-1B Specialty Occupations Visa Program, and organizations, such as O'Grady Peyton International, enable, incentivize, and reward health workers who leave their countries of origin to assume positions in the developed world. Notice that the movement of workers from developing countries to developed countries is not caused *only* by the choices of individual health workers; the movement of workers is *enabled* and *facilitated* through social structures that actively recruit these workers. While health workers exercise free will in deciding to emigrate, a web of relations among social positions (involving senators, hospital executives, recruiters, and others) makes such a choice possible.

This structure is vicious insofar as it actively recruits and grants visas to medical professionals from nations suffering from severe health workers shortages. The active recruiting of these workers emerges from a web of relations that harms the health and well-being of the poor, exacerbates global health injustices, and undermines the global common good. To be clear: I am not arguing that immigration is structurally vicious. Like the WHO, my position is not anti-immigration. With the WHO, I argue against the active recruiting and targeting of medical professionals from nations with low health worker densities.

In recognition of the harmful effects of recruiting, in May 2010 the WHO passed the Global Code of Practice on the International Recruitment of Health Personnel. The importance of the issue is evidenced by the fact that this is only the second Code approved by the WHO and the first one approved in twenty-nine years.[33] The Code's overarching goal is the ethical recruitment of health workers throughout the globe. In the absence of authentic global governance, such as treaties, the WHO encourages source and destination member states to develop non-binding, voluntary agreements to promote just recruiting practices.

---

[32] World Health Organization, "Global Code of Practice on the International Recruitment of Health Personnel," Preamble. Emphasis added.

[33] See World Health Organization, "User's Guide to the WHO Global Code of Practice on the International Recruitment of Health Personnel," www.who.int/hrh/resources/guide/en/index.html. The previous Code was: World Health Organization, "International Code of Marketing of Breast-Milk Substitutes," www.who.int/nutrition/publications/infantfeeding/9241541601/en/.

The Code directs destination nations to account for the health needs of the source nation when evaluating recruiting practices.[34] Article five is the centerpiece of the Code as it calls on developed nations to "discourage active recruitment of health personnel from developing countries facing critical shortages of health workers."[35] This prohibits recruiting from many Sub-Saharan and Southeast Asian nations.

However, as of the spring of 2020, the Code has been ineffective.[36] Consider the fact that while the Code prescribes monitoring and reporting practices for WHO members, only eighty of the 192-member states of the WHO have submitted reports concerning their progress toward implementing the Code. Only fifty-four of those eighty reporting-states have taken steps to implement the Code.[37] While the Code's effectiveness relies on voluntary bilateral and multilateral agreements between member states, by May of 2019, only seventy-seven agreements had been reached.

Because of the outsized role of Catholic-sponsored health care on the global stage, I offer three prescriptions for the Catholic church regarding the Code. First, the Catholic magisterium should publicly endorse the WHO's Global Code of Practice on the International Recruitment of Health Personnel. The issue is within the purview of the Pontifical Council for Justice and Peace and the Pontifical Council for Health Care Workers and, since 2017, of the Dicastery for Promoting Integral Human Development that assumed their competences. Furthermore, the US Catholic Bishops should revise their *Ethical and Religious Directives* to include prohibitions against recruiting in the Global South.[38] Because Catholic facilities contain fifteen percent of the hospital beds in the United States, and twenty-five percent of the beds in the world, action by the church will have a significant effect.[39]

---

[34] See World Health Organization, "Global Code of Practice on the International Recruitment of Health Personnel," par. 3.4.
[35] World Health Organization, "Global Code of Practice on the International Recruitment of Health Personnel," par. 5.1.
[36] See Vivian Tam et al., "Empirically Evaluating the WHO Global Code of Practice on the International Recruitment of Health Personnel's Impact on Four High-Income Countries Four Years after Adoption," *Global Health* 12, no. 1 (2016): doi.org/10.1186/s12992-016-0198-0.
[37] See World Health Organization, "Human Resources for Health," www.apps.who.int/gb/ebwha/pdf_files/WHA72/A72_23-en.pdf. Steps include laws, policies, and good practices that have been passed or are under consideration by member states and by the recruiting agencies within those states.
[38] See United States Conference of Catholic Bishops, *Ethical and Religious Directives for Catholic Health Care Services*, 6th ed. (Washington, DC: United States Conference of Catholic Bishops, 2018).
[39] See Catholic News Agency, "Catholic Hospitals Comprise One Quarter of World's Health Care, Council Reports," www.catholicnewsagency.com/news/catholic_

Second, Catholics should lobby for WHO member states and health care organizations to adopt and enforce the Global Code. Article five, if enforced, would stem the flow of workers from the neediest parts of the South. This recommendation is only politically feasible if the nations of the North commit resources to grow their own sustainable health workforces. Such growth can be accomplished. In 2000, the United Kingdom scaled-up domestic training programs for health workers and, by 2007, the country had dramatically reduced its reliance on recruiting foreign health workers.[40]

Third, both of the crises discussed in this chapter require the church to amplify its call for substantive global governance. Catholic social teaching has called for a robust global authority since St. John XXIII's 1962 social encyclical *Pacem in Terris*. Pope Benedict XVI continued this teaching in his 2009 encyclical *Caritas in Veritate* when he argued that a global authority must have "teeth" (no. 67). The Global Code is toothless and has been critiqued as such by traditional source nations such as Malawi. Voluntary codes are a start but will not guarantee that recruiting in the South will stop.

If enacted, these prescriptions would contribute to the creation of a structure of virtue—more specifically, a structure of solidarity—in which all peoples and nations would bear the burdens of the Global South and in which the global common good could emerge. While virtuous recruiting practices will not solve the health worker crises that plague parts of the Global South, such practices will be an essential step in broadening access to primary medical care.

Conclusion

This chapter demonstrates that an account of social structure facilitates ethical analyses of global warming and health worker shortages in the Global South. It also suggests that such an account could aid analyses in many other areas of global public health ethics. In the end, understanding social structures is necessary for global public health ethicists to render prudential claims that transform the structures that diminish and kill into structures that give life and life in abundance.[41]

Daniel J. Daly was associate professor of moral theology at Saint Anselm College in Manchester, New Hampshire, between 2008 and 2019. In 2008, he earned a PhD in Theological Ethics from Boston College. His research and

---

hospitals_represent_26_percent_of_worlds_health_facilities_reports_pontifical_council/.

[40] See Nigel Crisp, *Turning the World Upside Down: The Search for Global Health in the Twenty-First Century* (Boca Raton, FL: Taylor and Francis, 2010), 73.

[41] Special thanks to Christian Lingner for providing research and copy-editing assistance for this project.

publications fall into three categories. First, his current project develops a language for ethically scrutinizing global social structures to understand how social structures have a discernible moral character and to investigate how social structures facilitate or impede personal and social virtuous life. Second, he is interested in questions in fundamental Catholic theological ethics: from happiness and the virtues to natural law and moral norms, from the ethics of Thomas Aquinas to the common good. Third, by focusing on medical ethics and, in particular, on death and dying, he reflects on how the Catholic tradition can help in clinical decision making in an age of extraordinary medical resources and deep inequalities in access. He serves on the ethics boards of local hospitals. As a clinical medical ethicist, he assists in the adjudication of problematic cases that arise in these hospitals. In 2019, Prof. Daly joined the School of Theology and Ministry at Boston College. In 2020, he published the book *The Structures of Virtue and Vice*.

Ethics and Equity in Global Health: The Preferential Option for the Poor

Alexandre A. Martins

I BEGIN THIS CHAPTER BY CLEARLY STATING something that is not ideological but rather a fact: the main cause of health issues, diseases, and premature death is poverty.[1] Poverty creates a vicious cycle[2] that begins with injustice and ends with death. Poverty is not a natural phenomenon but a socioeconomic creation that makes people *vulnerable to fall ill*. Once sick, a poor person does *not have access to the medical care* needed to recover. This vicious circle leads to *more suffering* making people poorer and sicker. As a result, the poor *person dies* in a context that denies his/her dignity.

Moreover, I want to state a perspective that I believe is necessary to break this vicious cycle responsible for creating vulnerability for many victims of violence against their dignity. This is a perspective centered on the social *locus* where these victims of structural violence are; it is a perspective *from below*, from the experience of the poor, that places their voices at the center of our discussion and actions through a preferential option for the poor.[3]

---

[1] There are many comprehensive reports with studies showing the connection between poverty and illness. These reports are promoted by organizations, such as the World Health Organization (WHO), the World Bank, and the *Lancet* works teams, and can be found on their websites. See also Nazim Habibov et al., "Poverty Does Make Us Sick," *Annals of Global Health* 85, no. 1 (2019): 1–12.

[2] Norman Daniels et al. speak of a cycle where poverty causes bad health that makes vulnerable people even poorer, and thus produces a lower health status. See Norman Daniels et al., "Health and Inequality, or, Why Justice Is Good for Our Health," in *Public Health, Ethics, and Equity*, ed. Sudhir Anand et al. (Oxford: Oxford University Press, 2004), 63–92, at 65–66.

[3] The preferential option for the poor is a principle of Catholic social teaching that appeared for the first time in a Church document in 1987 in John Paul II's *Sollicitudo Rei Socialis*, no. 39. The Pope incorporated this principle in the official magisterium after its development in Latin American theology. Then, Pope Francis continued this tradition developing the meaning of the preferential option for the poor as a choice centered on Christological faith. See *Evangelii Gaudium*, no. 198, and *Laudato Si'*, no. 158. Although the preferential option for the poor has been broadly accepted in theological cycles around the world, its understanding focuses more on a conceptual-theoretical aspect than on its practical implications, as a way of life and as a

In this chapter, I will not be restricted to offering a theoretical discussion about the preferential option for the poor, as a perspective that has a huge potential to contribute to equity in global health. There are many books showing it.[4] Rather, I witness what the preferential option for the poor represents in practice. Hence, I begin by showing that this option is an existential commitment, an ethical imperative that guides a way of living and inspires decision-making processes. This chapter has two sections. First, I present the preferential option for the poor as an ethical imperative that sustains an existential commitment, with personal and professional implications. Second, considering the need of justice for health equity, I listen to some voices from the poor, allowing their experiences to engage us in searching for ways and practices to break the vicious cycle of poverty-vulnerability-lack of healthcare-premature death. This collective and mutual endeavor may help us to work for equity in health care.

OPTION FOR THE POOR: AN ETHICAL IMPERATIVE AND AN EXISTENTIAL COMMITMENT

As a Catholic bioethicist and theologian, I begin my reasoning and action with an encounter that demands a reaction or, in traditional and spiritual language, it begins with a call that requires an answer, what the ecclesial tradition names vocation. Essentially, such an encounter is with the face of the crucified Jesus—the face of a person who poured out his blood to love the dignity of the human being. This blood coming from the cross is pouring on the face of those who, throughout history, have their dignity crucified by violent and unjust forces. Borrowing Ignacio Ellacuría's terms, the encounter with the crucified Jesus—who in his suffering comforts us, as he comforted his disciple and mother close to his cross (John 19: 25–27)—demands a reaction of compassion that begins with a movement of seeing, a seeing the faces of those who are also suffering in history as victims of injustice. They are crucified people.[5] Therefore, the Latin American Bishops, in

---

perspective for building justice. This practical implication is only articulated in the original framing of the preferential option for the poor that was based on the embodied experience of Latin American communities in the 1950s, sixties, and seventies. To understand this process, see Alexandre A. Martins, *The Cry of the Poor: Liberation Ethics and Justice in Health Care* (Lanham, MD: Lexington Books, 2019), 59–75.

[4] See, for example, Gustavo Gutiérrez, "The Option for the Poor Arises from Faith in Christ," *Theological Studies* 70, no. 2 (2009): 317–326; Daniel G. Groody, and Gustavo Gutiérrez, ed., *The Preferential Option for the Poor Beyond Theology* (Notre Dame, IN: University of Notre Dame Press, 2014).

[5] See Ignacio Ellacuría, "The Crucified People: An Essay in Historical Soteriology," in *Ignacio Ellacuria: Essays on History, Liberation, and Salvation*, ed. Michael E. Lee (Maryknoll, NY: Orbis Books, 2013), 195–224.

the Conference of Aparecida (2007), affirmed: "The suffering faces of the poor are the suffering faces of Christ."[6]

Theologically speaking, the preferential option for the poor is a response to Jesus's call to be his disciples. This response creates a commitment that goes beyond any attempt to define a concept. As a commitment, the preferential option for the poor becomes an ethical principle that guides one's lifestyle and decisions with personal, social, and political implications. In his apostolic exhortation *Evangelii Gaudium*, Pope Francis provides this theological view: "The Church has made an option for the poor which is understood as a 'special form of primacy in the exercise of Christian charity'" (no. 198).[7] Moreover, he quotes from Pope Benedict XVI's address at the conference of Aparecida to affirm that the option for the poor depends on the Christological faith "in a God who became poor for us, so as to enrich us with his poverty."[8] Then Pope Francis continues: "I want a Church which is poor for the poor. They have much to teach us. Not only do they share in the *sensus fidei*, but in their difficulties they know the suffering Christ. We need to let ourselves be evangelized by them" (*Evangelii Gaudium*, no. 198). Being with the poor as an answer to Jesus's call is a commitment to a lifestyle that opens our ears to listen to the poor and learn from them.

Sociologically speaking, the preferential option for the poor leads us to where injustice occurs, to see the faces of the poor, listen to their voices, and learn from the stories of those who carry the burden of injustice and suffer its consequences, which often enough include death. In this reality, therefore, seeking for the causes of injustice and premature deaths becomes something natural. As an academic discipline, theology does not have authority *per se* to find these causes, but it joins sociological disciplines, with their methods and tools, to understand those causes and to find partners to address the socioeconomic roots of injustice.[9]

---

[6] "Los rostros sufrientes de los pobres son rostros sufrientes de Cristo." Consejo Episcopal Latinoamericano (CELAM), *V Conferencia General del Episcopado Latinoamericano y del Caribe, Documento Conclusivo, Aparecida, 13-31 de Mayo 2007*, 2nd ed. (Bogotá: Centro de Publicaciones del CELAM, 2007), no. 393. The Bishops quote Consejo Episcopal Latinoamericano (CELAM), *IV Conferencia General del Episcopado Latinoamericano y del Caribe, Documento Conclusivo, Santo Domingo, 12-28 de Octubre 1992*, 2nd ed. (Bogotá: Centro de Publicaciones del CELAM, 1992), no. 178.

[7] The quote is from John Paul II, *Sollicitudo Rei Socialis*, no. 42.

[8] Benedict XVI, "Address at the Inaugural Session of the Fifth General Conference of the Latin American and Caribbean Bishops," www.vatican.va/content/benedict-xvi/en/speeches/2007/may/documents/hf_ben-xvi_spe_20070513_conference-aparecida.html.

[9] In the experiences of Latin American communities, where liberation theology and the option for the poor originated, *a new way of doing theology* began with the step "0," that is, with joining the poor in their reality in an experience of community and

The preferential option for the poor is a perspective that creates partners who listen to one another and work together to address the roots of injustice. As a concept, the preferential option for the poor was first stated by the Latin American bishops in their regional conferences in 1968 and 1979,[10] but as an existential commitment, it always existed having the first example in Jesus of Nazareth. When Latin American Bishops and theologians began to write about the need for an option for the poor, many communities and committed people had already embodied this option. They were among the poor, acting with them or, to say it better, they were the poor. While the transformation of the option for the poor into a concept had its benefits, it also led the option for the poor to lose its existential aspect, having people who talk about the poor and the option for them but are not among the poor.[11] This is what I experience now in my own paradoxical existential drama. On the one hand, I can write and talk about the preferential option for the poor because it has become a theological concept. As such, scholars are interested in hearing about and studying it. On the other hand, I run the risk of talking about the poor while forgetting the poor, listening only to scholars and not to the voices of the poor.

To decrease this risk, Pope Francis made an important contribution. For Francis, we cannot talk substantially about the poor and not be committed to them in our way of living and in being with them. Thus, coherent with his own practice and with the Catholic tradition, Francis elevates the option for the poor to an ethical imperative that is necessary if the poor are to participate in the common good. He writes:

> The principle of the common good immediately becomes, logically and inevitably, a summons to solidarity and a preferential option for the poorest of our brothers and sisters. This option entails recognizing the implication of the universal destination of the world's goods, but ... it demands before all else an appreciation of the immense dignity of the poor in the light of our deepest convictions as believers. We need only look around us to see that, today, this option is in fact an ethical imperative for effectively attaining the common good (*Laudato Si'*, no. 158).

---

learning. See Leonardo Boff and Clodovis Boff, *Como Fazer Teologia da Libertação*, 8th ed. (Petrópolis: Vozes, 2001).

[10] The Conferences of Latin American and Caribbean Bishops of Medellín (1968) and Puebla (1979) embraced the preferential option for the poor that was being promoted by the basic ecclesial communities. See the documents of Medellín (no. 7) and Puebla (no. 1134), in Conselho Episcopal Latino-Americano, *Documentos do CELAM: Rio de Janeiro, Medellín, Puebla e Santo Domingo* (São Paulo: Paulus, 2005).

[11] See Martins, *The Cry of the Poor*, 59–75.

Francis believes that the market forces alone are not capable of promoting care able to answer the cry of the earth and of the poor (see *Laudato Si'*, nos. 49, 109, 190). Therefore, he argues for an integral perspective of development in which all participate, including the poor, who are not only recipients of actions but active agents of transformation (*Laudato Si'*, no. 179). Their voices matter. They have a knowledge that counts.

As an ethical imperative, the preferential option for the poor means allowing ourselves to be poor (whether spiritual and/or material poverty) to learn from the poor and their suffering, from their reality, and their beauty. The preferential option for the poor suggests an ethic of personal commitment and a collective effort. An option for the poor against poverty will break the cycle of violence against the dignity of human beings who are vulnerable and who suffer socioeconomic poverty.

JUSTICE FOR HEALTH EQUITY: THE VOICE AND THE EXPERIENCE OF THE POOR

Global public health has been neglected by bioethics, especially in the US, where bioethicists are very busy dealing with clinical issues and health professional-patient conflicts because of patients' autonomy and the use of new technologies.[12] As a consequence, *justice*—one of the principles of the Georgetown mantra (i.e., respect for autonomy, beneficence, nonmaleficence, and justice)[13]—is strongly approached as an individual perspective regarding those who have access to medical care. This approach, however, diminishes the social aspect of justice, limiting justice to what concerns the relationship between two individuals.

Global public health is about justice in healthcare and incorporates the participation of individuals in the common good. A justice that fosters participation in the common good results from a collective engagement that promotes fair relationships among individuals. Hence, justice creates conditions for people to participate in the common good, such as in healthcare, where individuals seeking care will find other individuals who will provide care, and their relationships will promote the well-being of all those involved.

In global health, the focus on justice might be so strong that sometimes one forgets the individual, especially humble persons who live in impoverished areas. It is not rare that global health initiatives are being developed in offices of affluent countries, influenced by the

---

[12] Paul Farmer, for example, is one global health scholar who presents this criticism. See Paul Farmer, *Pathologies of Power: Health, Human Rights, and the New War on the Poor* (Berkeley: University of California Press, 2003), 196–212.

[13] On these principles and what they mean, see Alastair V. Campbell, *Bioethics: The Basics* (London: Routledge, 2017), 44–47.

Western culture, and to be implemented in low-income regions, with totally different cultures and worldviews. The implementation of such projects can foster potential conflicts and their cultural fragility could lead to failure because "experts" ignore the individuals living in the context they want to serve. Often enough, these global health experts presume to know more about the life and experience of the poor than the poor themselves.[14]

The implementation of global health projects in local realities has a top-down approach, and the top-down approach in global health is part of the mainstream. Many organizations, universities, and governmental sponsored projects in this field rely on this approach. I do not question their good intentions, but I question their efficacy and their ethical respect for particularities and the self-determination of those people they aim to serve. At the 2019 conference *Ethical Challenges in Global Public Health* sponsored by Boston College,[15] presenters and participants agreed that global health actions must involve local communities, valuing their knowledge, culture, and experiences. Thus, global health initiatives promote actions with local partners in order to create independence.

However, an approach that begins from the bottom, empowering local communities, is not one that some leading global health organizations embrace, such as the World Health Organization. Thana Cristina de Campos, an expert in global bioethics and assistant professor at Pontificia Universidad Católica de Chile, stresses that centralization is the mainstream perspective in global health governance, with actions, projects, and systems being controlled by a central power—whether it is the WHO or the United Nations. According to Campos, lack of clear common ends as well as inclusion of local communities and coordination are the main problems in global health governance, recognized by those who defend centralization. However, centralization cannot properly address these problems. Thana de Campos argues for a global health governance grounded on the principle of subsidiarity.[16] This principle—which is articulated in Catholic social teaching and was included in the *Treaty on European Union* (Article 5)[17]—proposes a global health governance that includes local experiences and communities, empowering and respecting their own particularities. This empowerment would occur

---

[14] I can affirm it as someone who has been on both sides in different life contexts.
[15] See Schiller Institute for Integrated Science and Society at Boston College, "Ethical Challenges in Global Public Health," (2019), www.bc.edu/bc-web/academics/sites/ila/events/Ethical-changes-public-health.html#about_the_conference.
[16] See Thana Cristina de Campos, *The Global Health Crisis: Ethical Responsibilities* (Cambridge, UK: Cambridge University Press, 2017), 219–260.
[17] See Council of European Communities, *Treaty on European Union* (Luxembourg: Office for Official Publications of the European Communities, 1992).

because the principle of subsidiarity is based on three pillars: non-abandonment, non-absorption, and coordination. These three aspects are more appropriate to address the three main challenges of global health governance in ways that empower communities as agents of global health by promoting their own reality, including their worldview.

Thana de Campos's suggestion of relying on the principle of subsidiarity to shape global health governance shares an approach to global health promotion grounded on the preferential option for the poor. This option directs global health initiatives to engage local communities, considering the knowledge of locals and their experience in the midst of poverty. Therefore, the poor become active partners of global health and not only recipients of charitable actions from affluent nations.

Because justice and participation in the common good are central ethical challenges in global health, the preferential option for the poor offers a perspective that inverts the most common approach to health care by listening to the voices of the poor and by engaging the poor as core partners in the effort of promoting global health. This approach breaks the vicious cycle created by impoverishment (poverty-vulnerability-illness-lack of health care-premature death). Moreover, such an approach exemplifies a perspective *from below* that respects the particularities of local communities because the voices of the poor matter. Hence, it can address the common criticism that global health organizations promote a new form of colonialism because they create dependency and do not work for the empowerment of impoverished communities and their development.

Moreover, equity in global health is an ethical issue because, at its center, it concerns justice: *social justice*—structures and services that can create access to health care—and *justice as empowerment* of local people, by promoting their participation with their own unique contributions, in decision-making processes. These are key elements to foster social justice and democracy.

As an ethical imperative and existential commitment, the preferential option for the poor has a pragmatic aspect that unites social justice and justice as empowerment. To use Paul Farmer's words, the preferential option for the poor suggests a "pragmatic solidarity"[18] that orients: (1) global health initiatives and systems with global, public structural impacts—such as public health programs that fill unjust gaps, giving the poor priority to create health equity; and (2) local actions and interactions by listening to the poor in processes regarding health decisions and promotion. In the case of local actions and interactions, the preferential option for the poor leads us to create

---

[18] Farmer, *Pathologies of Power*, 26.

a *process of mutual learning*[19] between the knowledge of the poor and the expertise of professionals and global public health advocates.

Being with the poor implies compassion, listening to them, in order to understand their suffering and reality. From the places where I have worked, such as Bolivia and Uganda, I bring some of their voices that we should hear, while we feel their suffering. Such listening is like an exercise of contemplation, thinking about their crucified faces, and how we can engage with these people, sharing our technical knowledge while we learn from their lives. This exercise creates a process of mutual learning and promotes a prophetic praxis.

From Bolivia:
I was feeling very bad, in a way I couldn't walk. Then I passed out. My neighbor came to my home, saw I was out and took me to the hospital. Doctors helped me as much as they could. Then they told me that I passed out because my diabetes was so high and that I needed insulin. They told me "we gave insulin to you now, but you have to pay for that and if you don't buy more, you can't apply anymore in you." I didn't have money that time and my neighbor paid for me. Then my family came and bought more (Madalena).

Doctors do not give much attention, especially if you are poor and come from the countryside. As they don't provide exams and medication, we have to buy or go to the particular [a private clinic]. When you pay, the doctor treats you well (Raimunda).

The community helps. In the neighborhood, there are people who organize a "quermese solidaria" [an event of solidarity], as we call here, to help with the cost that he has [cost of medical treatment]. But we can't pay for everything. Well, people have solidarity, raise money, but it is not enough in many cases because [medical] materials are very expensive. The treatment that one has to do is very expensive and our community can't raise enough money. Bolivians are very hospitable; we are united. The unique problem is that many people can't help because they don't have money (Josephina).

From Uganda:
I am a Buganda person, and I don't think like you. Foreigners and missionaries come here to help us, like you. They do a good work, but they do not trust us. They treat us as

---

[19] See Martins, *The Cry of the Poor*, 214–216.

if we are incapable of doing anything right. All the good work is attached to foreign missionaries and their international groups of people. They make us to be dependent on them and accept everything they are saying. We are helped when we are sick, in our immediate need, but we never advance to an independent life and to grow in a Buganda way (Archelo).

Please, don't take pictures of us. Mzungu comes here, take pictures with us, especially with children on street. Then Mzungu shows this picture in Europe and United State to make money. Rich people will give them money and this money never will come here to help us (Kyakuwaile).

These voices show their experience, dramas, and struggle. They also reveal their creativity and their view of top-down actions. Engaging with the poor promotes a process of transformation with mutual collaboration. As I write in another work, where the concept of mutual learning is developed after working with the poor, "The voices of the poor will tell us what their suffering and struggle mean, and help us shape collective actions able to respond to their social and health inequality challenges, without an anthropological destruction."[20]

CONCLUSION

To include the poor and their voices in a shared promotion of the common good, global health has a lot to learn *from* and *with* local communities, empowering their experiences and supporting their agency. Some global health leaders seem not to be in favor of this way of action. These people still believe that they have the authority to promote global health action by going to an impoverished region and simply acting for (and not with) the poor. No doubt to work *with* the poor requires more discussion and studies. The inclusion of ethics in the global health agenda and of global health in bioethical discussions will contribute to developing global health, especially in delivering health care in impoverished, diverse communities around the world.

Because justice is a central ethical challenge in global health, the preferential option for the poor offers a perspective that inverts the most common approach to health care, centered on well-off single individuals, by acknowledging the poor, with their voices, as core partners in the effort of promoting global health. We cannot change the world alone. Perhaps, we cannot significantly change the world of healthcare, but if we want to work for this change, we must begin with those who are the first ones to call for this change in their own reality,

---

[20] Martins, *The Cry of the Poor*, 128.

in a process of mutual learning, by breaking the vicious cycle caused by poverty. M

Alexandre A. Martins is assistant professor in the Department of Theology at Marquette University where he earned his PhD in 2017. He specializes in health care ethics and social ethics, especially in the areas of public health, global health, Catholic social teaching, and liberation theology. He is also a scholar in philosophy of religion, specialized in the work of the French philosopher Simone Weil. He has engaged ethics, theology, and healthcare in dialogue with anthropology, philosophy, epidemiology, and medical science, by focusing on marginalized voices and addressing issues from the perspective of the poor in their marginalized communities. He advised the Brazilian Bishops' Conference on issues of bioethics and pastoral care. With humanitarian organizations and in academic projects, he served in areas marked by poverty and lack of adequate health care assistance, such as Brazil, Bolivia, and Haiti. Widely published, he has lectured in various countries. He was a visiting scholar at Saint Camillus University in São Paulo, Brazil, and at the Camillianum–International Institute of Pastoral Theology in Health at the Pontifical Lateran University in Rome. He is also member of the Latin American Regional Committee of the Catholic Theological Ethics in the World Church and of the Brazilian Society of Moral Theology.

Social Justice and the Common Good: Improving the Catholic Social Teaching Framework

Lisa Sowle Cahill

P**OPE FRANCIS'S ENCYCLICAL LETTER ON THE** environment, *Laudato Si'*, is a remarkable intervention into Catholic social teaching (CST). Its vision of environmental justice is groundbreaking. *Laudato Si'* eloquently proclaims the beauty of the natural universe and strikingly redefines humanity's place within it, by insisting that all creatures have intrinsic value and, with humanity, are included in the redeeming grace of Christ. Moreover, this encyclical and its successor, the 2019 Synod on the Amazon, provide more than an environmental ethics. Their implications for the CST framework as a whole are radical.

This chapter briefly recapitulates the modern history of CST, then situates three innovations of Pope Francis's ecological agenda in terms of their significance for the evolving tradition. All three derive from Pope Francis's indictment of powerful elites for obstructing international agreements on climate change and his tacit recognition that moral teaching alone cannot reverse standing injustices. I argue, however, that there is a significant blind spot in CST on the environment and health justice, one that even Pope Francis has still to address: gender equality. The chapter concludes with a case study from the Amazon that draws these themes together.

CATHOLIC SOCIAL TEACHING IN HISTORICAL PERSPECTIVE

CST has been promoting the common good as an indispensable criterion of social justice since at least the end of the nineteenth century, with the publication of the first of the modern papal social encyclicals, Leo XIII's *Rerum Novarum* (1893). CST has since expanded the concept's sphere of reference from the nation state and its constitutive communities (e.g., families, neighborhoods, towns, cities, provinces, states, and regions), to the "universal common good" of all nations and peoples; and, with Pope Francis, to that of the entire planet: "everything is connected," the recurring refrain of *Laudato Si'*.

One might even say that the common good *is* the definition of social justice in CST. As a definition or standard of justice, the common good entails the equal participation of every member of society in basic material, social, and political goods, both as a contributor and as a beneficiary. In fact, the common good is a form of a dynamic community that is a good in itself.

Additional bedrock concepts of CST are mutual rights and duties; solidarity as an active commitment to justice; good government, the rule of law, and just, participatory political institutions; and the principle of subsidiarity. This principle provides both that higher authorities do not interfere unduly with arrangements at the local level and that they do intervene when local institutions cannot or do not fulfill the requirements of justice. Importantly, responsibility for the common good includes not only civil government but also civil society, private institutions, and religious bodies. Somewhat paradoxically, however, the authors of the modern social encyclicals have often seemed to want or hope that the United Nations or a like body will step in to handle structural injustices, such as war and violence, economic inequality and development, human rights, and now ecological harm. As we will see, recent assessments of contemporary global governance structures have cast serious doubt on the ability of the UN to control the conditions of global justice as well as on the assumption that, if it did so, the results necessarily would be benign.

Most importantly, however, papal thought since the Second Vatican Council (1960–1965) has moved decisively in the direction of liberation theology's "preferential option for the poor" to define what the common good means concretely. This phrase makes the needs of the most marginal or vulnerable the ethical priority, as inspired by the gospel and Jesus's care for "the least of these" (Matthew 25). In justice terms, the preferential option for the poor means something like urgent affirmative action to improve the situation of the least well-off. In the name of the preferential option for the poor, Pope John Paul II urged that the gospel and justice both require the transformation of a world in which poverty is assuming "massive proportions" (*Centesimus Annus*, no. 57). Pope Francis has applied the preferential option not only to disenfranchised populations most vulnerable to climate change but to the Earth itself and all creatures harmed by selfish interests. In other words, liberal equality, procedural justice, and "equal opportunity" that prescinds from the real social, cultural, and economic conditions that limit individual choice and agency are not sufficient conditions of justice for CST. CST defines justice not only in terms of individual freedoms and rights, but also in terms of structural justice that respects the social, political, and material needs of all.

Alarmingly, however, over a century and a quarter of papal, episcopal, and theological teaching on these priorities—the common good, justice, and the priority of the poor in morality and politics—have done little to roll back injustice. Poverty, violence, prejudice, and exploitation are still around and still perpetrating their devastating effects on the world's least powerful peoples and their habitats. According to the United Nations,

> In 2015, more than 736 million people lived below the international poverty line. Around 10 per cent of the world population is living in extreme poverty and struggling to fulfil the most basic needs like health, education, and access to water and sanitation, to name a few. There are 122 women aged 25 to 34 living in poverty for every 100 men of the same age group, and more than 160 million children are at risk of continuing to live in extreme poverty by 2030.[1]

Social justice and the common good have obvious relevance to the problem of public health ethics, as does the reality of world poverty. From the standpoint of public health, "social justice" requires equity in access to health resources and to the social determinants of good health. The concept of the "common good" captures the *public and social* nature of health and health justice. Relatedly, the natural environment, now undermined by climate change, is a public good, one which demands not only a social but a cross-cultural and intergenerational understanding of justice. This is essential in order to address the effects of climate change and other environmental damage on health, especially for the planet's most vulnerable people. Particularly to be accented in defining and applying the concepts of social justice and the common good in a global public health environment is the participation of affected populations themselves, especially women.

In the twenty-first century, CST has major challenges to meet if it is to realize its ideals concretely. The top-down teaching approach and the social model based on the modern Western nation-state (i.e., rights, duties, democratic government, and rule of law) will no longer do. Both the theology and the politics of CST must be refocused if they are to have any prospect of transforming conditions of injustice more successfully than in the past; the voices and sources on which it draws must be much more local, bottom-up, and diverse.

Three main challenges the common good tradition now faces are the absence of real *political will* to make justice-favoring changes in society, the economy, or politics; the need to go beyond reason and exhortation to *conversion* of imaginations and worldviews; and the *decentralized* nature of global socioeconomic and political agency,

---

[1] United Nations, "Ending Poverty," www.un.org/en/sections/issues-depth/poverty/.

bringing with it popular mobilization as a prerequisite of accomplishing the environmental goals the Pope sets out. Moreover, special attention must go to the secondary status of women in virtually every society historically and today, contrasted to the reality of women's environmental activism. After elaborating on these challenges, I turn to a case from the Peruvian Amazon to illustrate the changed and evolving environment of CST.

THE CHALLENGE OF POLITICAL WILL

Now to challenge number one. *Laudato Si'*—unlike its predecessors—specifically recognizes that efforts to institutionalize justice internationally will always be opposed, usually effectively so, by powerful interests that are vested in the status quo. Civil society organizations have made some progress in creating an ecological movement. However, UN summits on the environment have not lived up to expectations because even when they succeed in producing agreements, they fail to be effective. As Pope Francis incisively puts it, this failure is due simply to a "lack of political will" (*Laudato Si'*, no. 166). Indeed, "It is remarkable how weak international political responses have been .... There are too many special interests, and economic interests easily end up trumping the common good...." (*Laudato Si'*, no. 54).

In 2019, four years after the Paris Climate Accord, US President Donald J. Trump initiated a process of withdrawal—now reversed by President Joseph Biden—many countries are failing to meet their pledges, and target goals are being reduced, while the window of opportunity for meaningful action is becoming narrower, the urgency of reform more acute, and the effects of delay on the global poor ever more dire. An example related to pollution and health is the extraction of natural resources in the Global South, especially mineral ore. A 2017 report of the UN Environment Programme called for mining corporations globally to take a "safety first" approach to the disposal of toxic waste. Yet, two years later, the UN Sustainable Development Report 2019 asserted that environmental decline and social inequities are worsening, due in no small part to consumption patterns that result in toxic effects from mining and other pollution sources. The report itself insists that transformations can come about only through "coordinated action by governments, business, communities, civil society and individuals."[2]

---

[2] "Scientists call for urgent, targeted action to avoid reversing the development gains of recent decades." UN Department of Economic and Social Affairs, "Global Sustainable Development Report 2019," www.un.org/development/desa/en/news/sustainable/global-sustainable-development-report-2019.html. See also UN Environment Program, "New Report Urges Global Action on Mining Pollution,"

The reality is that reasonable arguments, moral appeals, and even authoritative religious teaching will not be enough to penetrate the recalcitrant coalition of the threatened and the apathetic. A key ingredient of real change has to be mobilization from below, gathering enough momentum to successfully pressure the beneficiaries and guardians of major social institutions so that they go beyond the hypocrisy of unenforced accords and achieve some meaningful compliance with the common good (*Laudato Si'*, nos. 169, 180–181).

THE CHALLENGE OF MORAL AND POLITICAL CONVERSION

This brings us to challenge number two: conversion and empowerment. First and most obviously, let us return to *Laudato Si'*. Pope Francis invokes the social virtue of solidarity to prioritize consequences of environmental damage for the poor. "Obstructionist attitudes, even on the part of believers, can range from denial of the problem to indifference, nonchalant resignation or blind confidence in technical solutions. We require a new and universal solidarity" (*Laudato Si'*, no. 14). Yet, the strongest asset of the Pope's message, in my mind, is that he neither leaves solidarity at the ideal level nor assumes that moral appeals and reason will bring the necessary response. In addition and more profoundly, there must be a conversion of worldviews, imaginations, and spiritualities. For this, aesthetic and affective appeals are required, along with formation in new sorts of moral and social relationships, and a renewal of Christian narratives, symbols, and liturgies. This is the first encyclical that I know of to be released with a Vatican YouTube video that stirs the imagination with the wonders of the natural world, the horrors of drought and pollution, and their effects on creation's human inhabitants.[3] The encyclical stirringly depicts for us the beauty and suffering of "sister earth" and that of our fellow creatures, while proclaiming the presence of the divine in and through creation. It concludes with two prayers, one for Christians, another for all who believe in a Creator God.

Conversion of imaginations and worldviews becomes more specific in its sources and appeals with the preparatory and final documents of the Amazon Synod and in the post-synodal Apostolic Exhortation *Querida Amazonia*. The preparatory document for the Amazon Synod leads the way by emphasizing even more strongly than *Laudato Si'* that the peoples most affected by climate change are also the most competent to diagnose the damage and to propose solutions. The peoples of the Amazon, including indigenous peoples and religions, have a cosmological and spiritual vision that must be shared

---

www.unenvironment.org/news-and-stories/story/new-report-urges-global-action-mining-pollution.

[3] See United States Conference of Catholic Bishops, "Vatican Releases Video on Pope's Encyclical," www.youtube.com/watch?v=KXA5_juFgDg.

if those in the Global North, responsible for most of the Amazon's ecological woes, are to imagine a different way of acting on the idea that we are all connected. In *Querida Amazonia*, Pope Francis very personally shares his four "dreams" for the Amazon, including the rights, dignity, and voices of the original peoples; the preservation of the region's "cultural riches" that reveal the varied "beauty of our humanity"; the preservation of the Amazon's "overwhelming natural beauty and the superabundant life teeming in its rivers and forests"; and of a Church with "new faces with Amazonian features."[4]

THE CHALLENGE OF A "NEW WORLD ORDER"

The opening plan for the Amazon asserted that justice for the region "requires structural and personal changes by all human beings, by nations, and by the Church."[5] A relatively recent political science literature contests CST's traditional top-down framework, showing not only that the United Nations has in reality very little enforcement power but also that some of its more momentous decisions depend on the agreement of its Security Council, which will always be made up of states with their own interests, biases, and disagreements. UN resolutions or treaties—on climate change, mining, or development goals—still have significant moral authority. Yet their real impact depends on implementation back home by member states, where political support may be low and resistance from special interests high.

In regard to global economic and political agency, political scientist Anne Marie Slaughter calls attention to what she calls a twenty-first century "new world order." Slaughter and others maintain that, over the past few decades, global economic and political power has become less and less centralized.[6] Successful regulatory, legislative, and judicial approaches of individual states can attract by 'soft power', prompting adaptation by other states and creating a *de facto* transnational policy regime upheld by like-minded counterparts, though not structured or enforced by any higher political power. These emerging networks can both obstruct international environmental target plans and provide new axes along which to garner support for their goals.[7]

---

[4] Francis, "Post-Synodal Apostolic Exhortation *Querida Amazonia*," www.vatican.va/content/francesco/en/apost_exhortations/documents/papa-francesco_esortazione-ap_20200202_querida-amazonia.html, no. 7.
[5] Synod of Bishops of the Pan-Amazon Region, "Preparatory Document for the Synod on the Pan-Amazon Region," www.sinodoamazonico.va/content/sinodoamazonico/en/documents/preparatory-document-for-the-synod-for-the-amazon.html, Preamble.
[6] See Anne-Marie Slaughter, *A New World Order* (Princeton, NJ: Princeton University Press, 2004); Maryann Cusimano Love, *Beyond Sovereignty: Issues for a Global Agenda*, 4th ed. (Boston: Wadsworth, 2011).
[7] For a more cautious reading, see Stewart Patrick, "The Unruled World: The Case for Good Enough Global Governance," *Foreign Affairs* 93, no. 1 (2014): 58–73.

Pope Francis grasps, in a way revolutionary for CST, that effective political action must be broad-based and multi-layered, gathering energy and strength among affected populations. Good faith efforts by governments, regulatory bodies, and businesses are optimal. But citizen pressure, protests, boycotts, shareholder dissent, and socially responsible entrepreneurial competition all have an important role. In the United States, the Catholic Climate Covenant is a cross-society community organizing effort, joined by eighteen partners, including the United States Conference of Catholic Bishops (USCCB), Catholic Relief Services, Catholic Charities USA, the Catholic Health Association, and congregations of religious men and women.[8] In *Laudato Si'*, Francis addresses at least seventeen local bishops' conferences, and draws widely on local examples and wisdom. The four dreams of *Querida Amazonia* are addressed to the region itself, to the Amazonian church, and especially to the experiences, worldviews, and power of the peoples that inhabit it.

One of the first public, political Catholic responses to *Laudato Si'* was the mobilization of the Catholic Church in the Philippines, among those nations most vulnerable to climate insecurity. Within a month of the encyclical's publication, Cardinal Luis Antonio Tagle of Manila launched a campaign to collect one million signatures on a petition to be delivered to Paris in December 2015 for the UN climate conference.[9] In July 2019, the Filipino bishops issued a pastoral letter, in which they urge that financial resources of Catholic institutions be disinvested from "dirty energy" like "coal-fired power plants, mining companies and other destructive extractive projects," and reinvested in clean and renewable energy sources.[10] To bring lasting change, popular momentum must meet up with policy and regulation, as well as conversion or motivation of the plants, companies, and other beneficiaries behind the "destructive" projects responsible for climate change and pollution. In a case study to follow, I show that various forces favoring environmental justice in the Amazon can converge and what sorts of barriers they still face.

---

[8] See Catholic Climate Covenant, "Our Story," catholicclimatecovenant.org/about/story.

[9] See Brian Roewe, "Philippine Church Takes Lead on Pope Francis's Climate Encyclical," *National Catholic Reporter*, July 25, 2015, www.ncronline.org/blogs/eco-catholic/philippine-church-takes-lead-francis-environmental-encyclical.

[10] Robin Gomes, "Filipino Bishops Issue Pastoral Letter on 'Climate Emergency'," *Vatican News*, July 16, 2019, www.vaticannews.va/en/church/news/2019-07/philippines-bishops-cbcp-pastoral-letter-climate-emergency.html.

CATHOLIC SOCIAL TEACHING, GENDER EQUALITY, AND WOMEN'S AGENCY

A key dimension of just local empowerment is *gender equality*. Despite my admiration for *Laudato Si'*, one cannot avoid the conclusion that it, like subsequent Amazon documents, shares a blind spot with CST as a whole and with many other faith-based advocacy efforts, and that is gender equality.[11] Just as in situations of civil conflict and war, it is often women who are most active in energizing solutions to climate threats at the grassroots level, perhaps because women are typically most responsible for the daily sustenance and safety of families and homes. In fact, the *Laudato Si'* video captures this with images of women in Africa waiting in lines to carry scarce water back for the domestic daily supply.[12] Yet the encyclical portrays our sister earth as a victim of human exploitation who awaits rescue at human hands, without noticing that the hands cultivating crops or finding water solutions worldwide are likely as not those of women. The gender symbolism of the encyclical places women in a dependent situation of victimization, while suggesting that women can be defined by familial, especially maternal, roles.[13]

Meanwhile, women's actual efforts to combat climate change and advocate for the health of families and communities are key. In fact, despite cultural and ecclesial gender barriers, women are leaders in mining resistance movements.[14] Yet, because Amazonian women must construct active political roles within highly gendered social norms reinforced by religion and culture, they usually meet greater difficulty than men in accommodating activism to their other roles, especially to domestic responsibilities. They also meet greater and more violent resistance when transgressing perceived norms. This includes domestic abuse and physical attacks, extending to rape and murder. When documents of Catholic social teaching continue to

---

[11] On gender in *Laudato Si'*, see Emily Reimer-Barry, "On Naming God: Gendered God-Talk in *Laudato Si'*," *Catholic Moral Theology*, catholicmoraltheology.com/on-naming-god-gendered-god-talk-in-laudato-si/; Nichole M. Flores, "'Our Sister, Mother Earth': Solidarity and Familial Ecology in *Laudato Si'*," *Journal of Religious Ethics* 46, no. 3 (2018): 463–478; Meghan Clark, "'Querida Amazonia' Provides Reflection on Ecological Sin," *National Catholic Reporter*, February 14, 2020, www.ncronline.org/news/earthbeat/querida-amazonia-provides-reflection-ecological-sin.

[12] See United States Conference of Catholic Bishops, "Vatican Releases Video on Pope's Encyclical."

[13] See Tina Beattie, "A 'Frozen' Idea of the Feminine," *The Tablet*, February 20, 2020, www.thetablet.co.uk/features/2/17583/a-frozen-idea-of-the-feminine.

[14] See Katy Jenkins, "Unearthing Women's Anti-Mining Activism in the Andes: Pachamama and the 'Mad Old Women'," *Antipode* 47, no. 2 (2015): 442–460; Katy Jenkins, "Women Anti-Mining Activists' Narratives of Everyday Resistance in the Andes: Staying Put and Carrying on in Peru and Ecuador," *Gender Place and Culture* 24, no. 10 (2017): 1441–1459.

reiterate traditional stereotypes of women's nature and roles, while ostensibly aiming to "respect" and "protect" women, they fail to empower women's environmental agency and women's ability to take initiative in respecting and protecting "our common home," as *Laudato Si'* urges. Traditionalist Catholic gender norms, when co-opted by violent political and economic forces within patriarchal cultures (virtually all cultures), can contribute to retaliation against women political activists. Women's work needs and deserves recognition and support from Catholic teaching, Catholic organizations, Catholic leaders, and Catholic co-workers.

The Final Document for the Amazonian Synod promisingly holds up the empowering example of Mary Magdalene, the "apostle to the apostles," as a model for the church in the Amazon and potentially for Amazonian women.[15] But Pope Francis's subsequent Apostolic Exhortation, *Querida Amazonia*, despite its powerful testimony to the wisdom of indigenous peoples, worldviews, and spiritualities, seems to undercut women's full participation. Women's positions and service in the Church should be respected, Francis says, but our gifts are "simple and straightforward" (more so than men's, apparently).[16] Women's work is approved when it "reflects their [simple] womanhood."[17] This criterion does not do justice to what women are actually doing to combat environmental destruction, as scientists, theologians, political representatives, NGOs officials, and community activists.

CASE STUDY: MINING IN MADRE DE DIOS, PERU

Let me conclude with a case bringing many of these themes together: the reduction and reform of gold mining in the Madre de Dios region of the Peruvian Amazon, one of the largest gold mining sites in the world. When national governments in the Amazon region grant licenses to international mining companies, they often displace indigenous peoples from ancestral lands, damage forests, and leave toxic waste materials that affect the health of whole populations for generations. Mining in the Amazon is connected to North American (especially Canadian) corporations and consumers, because metals including gold and tin are used to make electronic devices like smartphones.[18]

---

[15] See Synod of Bishops of the Pan-Amazon Region, "Final Document of the Amazon Synod: The Amazon: New Paths for the Church and for an Integral Ecology," www.synod.va/content/sinodoamazonico/en/documents/final-document-of-the-amazon-synod.html, at no. 22.
[16] Francis, *"Querida Amazonia,"* no. 102.
[17] Francis, *"Querida Amazonia,"* no. 103.
[18] See Kevin Jackson, "How the Amazon Synod Is Connected to North American Corporations," *America*, October 22, 2019, www.americamagazine.org/politics-society/2019/10/22/how-amazon-synod-connected-north-american-corporations.

There are many resulting problems, among which the following five are central.
1. Displacement of farmers. The Peruvian government is making profitable contracts with international mining companies, granting them rights to indigenous lands.
2. Extensive deforestation by the mining operations. Deforestation brings global damage, since the Amazon forests absorb a high amount of the world's carbon dioxide.
3. Toxicity from the mercury used to purify ore. Waste is dumped into the river or left to seep into the land, affecting food and water supplies.
4. Labor trafficking and sex trafficking. The economic profits and atmosphere of legal impunity around mines enables and conceals profitable human rights abuses.
5. Instability of employment and other means of income, especially agriculture. Eventually gold will be exhausted, land and water will have been polluted, and farming land destroyed.

Despite clear injustices and harm to the common good, political and economic incentives are on the side of continued exploitation. Local governments may enact some restrictions on mining but enforcement is weak. Meanwhile international accords, such as the 2007 UN International Declaration on the Rights of Indigenous Peoples, lack clear enforcement mechanisms, and the same political will to implement that is absent from agreements on climate change is lacking here as well.

Advocacy efforts and reform initiatives are active and growing, however. Perhaps most relevant and salient in relation to CST is REPAM (Red Eclesial Panamazónica), which organized the Synod on the Amazon. REPAM is a project of the nine Churches of the Amazon region, inspired by Pope Francis and backed by the Latin American Bishops' Conference, CELAM. Caritas Internationalis is a founding member of REPAM, and supporters of REPAM include national Caritas offices in the Amazon countries and Europe, as well as Catholic Relief Services (CRS) in North America. Catholic Relief Services' Madre de Dios area project manager is a woman, Tatiana Cottrina. The work of CRS in Peru is supported partly with a grant from the United States Agency for International Development, indicating the reach of Catholic international organizations.

Among other change agents in the Amazon region are the union of indigenous communities (Native Federation of Madre de Dios) and local activists like Nasbat Marleni who is a traditional cacao farmer who helped form an agroforestry organization and Victor Zambano who returned to his home in Madre de Dios after military service and has spent the last thirty years reforesting 84 acres of land.[19] Also involved are Peruvian scientists and organizations like the Amazon Conservation Association and the Center for Amazonian Scientific Innovation. Scientists have come up with a new technology for mining—a water table to sort gold—which leaves no toxic waste because it does not use mercury.

The Peruvian bishops and regional Amazonian bishops are helping to develop "Plans de Vida" by and for local communities with the support of Catholic organizations like CRS. Projects include fish farming, banana groves, ecotourism, crafts, reforestation, and agroforestry projects. The advice of Joseph Kelly, head of CRS operations in South America, resonates with CST:

> Start at the local level. Empower communities to have a voice in their environment, in how their economy develops and constructively build a way forward. Communities shift that paradigm and say, "We have a part to play. We have a voice. We have a role in changing this situation and we're protagonists in this story."[20]

In September 2019, over two dozen indigenous women leaders from across the Madre de Dios region met in the regional capital of Puerto Maldonado to unite against deforestation, extractive industries, and domestic violence and to learn from experts such as human rights lawyers and land rights defenders. One of the leaders, Jackelyn Rengifo, thirty years old, is president of the San Jacinto indigenous community in the Tambopata district of the Peruvian Amazon. Her community has been ravaged by the incursions of illegal and legal miners who have irreversibly damaged traditional lands. The community has a forest recovery project that includes growing fruits like oranges, tangerines, and cocoa. Inside the abandoned and once-toxic mining pits, they are aiming to construct fish farms.[21]

In the Madre de Dios region of the Peruvian Amazon, social justice and the common good are being realized ever more concretely with increasingly equitable participation by the most vulnerable stakeholders. New to the picture—and critically important for the climate and public health challenges to be faced in this century—are the empowerment of the vulnerable by motivated allies; their claiming of their *own* voice, wisdom and agency, especially by women; and the stirrings of local, regional, transnational, and global networks of

---

[19] See Rebekah Kates Lemske, "The Pull of Peru's Gold Rush," *CRS website*, www.crs.org/stories/peru-illegal-gold-mining-climate-change.
[20] Lemske, "The Pull of Peru's Gold Rush."
[21] See Andrew J. Wright, "Indigenous Women in the Peruvian Amazon Are Leading the Fight for Rights," *Sojourners Magazine*, September 18, 2019, sojo.net/articles/indigenous-women-peruvian-amazon-are-leading-fight-rights.

action committed to progress toward greater environmental and public health justice. 🅼

Lisa Sowle Cahill received her BA in Theology from Santa Clara University in 1970, followed by MA and PhD degrees from the University of Chicago Divinity School, where she wrote her dissertation under the supervision of James M. Gustafson. She is J. Donald Monan Professor in the Department of Theology at Boston College, where she has taught since 1976. She has been a Visiting Scholar at the Kennedy Institute of Ethics, Georgetown University, Yale University, and Dharmaram University in Bangalore, India. Dr. Cahill is a fellow of the American Academy of Arts and Sciences and has held office in the American Academy of Religion. She was president of the Catholic Theological Society of America and of the Society of Christian Ethics. She served as an editor, and on the editorial boards, of many prestigious journals. In addition, she has been a member of the Catholic Health Association Theology and Ethics Advisory Committee, the National Advisory Board for Ethics in Reproduction, and served on the March of Dimes National Bioethics Committee. She has given testimony to the National Bioethics Advisory Commission on fetal tissue research and on cloning. Her areas of interest and expertise, and very extensive publications, comprise the whole field of theological ethics. She has written or edited sixteen books and she is the author of over two hundred essays that appeared in books or journals.

Part 4:
International Approaches to Global Public Health

Challenges in Global Health, Culture, and Ethics in Africa

Jacquineau Azetsop, SJ

GROUNDED IN THE UNDERSTANDING OF HEALTH as a public good and in a larger vision of human community, citing Jeffrey P. Koplan and colleagues, Catherine Myser affirms that global health is

> an area for study, research, and practice that places a priority on improving health and achieving equity for all people worldwide ... [emphasizing] transnational health issues, determinants, and solutions; [involving] many disciplines within and beyond the health sciences and [promoting] interdisciplinary collaboration; and [synthesizing] population-based prevention with individual-level clinical care.[1]

From this definition, we can infer that global health is premised on a vision of human community grounded in social equity and our shared humanity. Therefore, social justice in global health cannot be merely seen as a second-level intellectual inquiry that aims at judging actions and practices performed by global health practitioners. Instead, social ethics is an integral part of global health vision that serves both as an instance for the conception and implementation of global health policy and as an entity that critically evaluates intervention outcomes.

Social justice requires that every person is entitled equally to key ends such as health protection and acquisition of the basic dimension of well-being.[2] Used as a framework for global health activism and policy, the concept of social justice broadens our moral horizon and includes many more interlocutors into the global health debate in such

---

[1] Catherine Myser, "Defining 'Global Health Ethics': Offering a Research Agenda for More Bioethics and Multidisciplinary Contributions—from the Global South and Beyond the Health Sciences—to Enrich Global Health and Global Health Ethics Initiatives," *Journal of Bioethical Inquiry* 12, no. 1 (2015): 5–10, at 6.

[2] See Madison Powers and Ruth R. Faden, *Social Justice: The Moral Foundations of Public Health and Health Policy* (Oxford: Oxford University Press, 2006), 1–14.

a way that global and local social challenges are shown to be connected through health challenges. Hence, the combination of health and social problems brings about structural questions that need to be confronted to ensure lasting health gains from global health interventions. Having said this, we cannot help but bring to the forefront three major challenges that need to be faced to ensure lasting health gains in most African countries: firstly, the fragmentation and pitiful shape of the healthcare system; secondly, the lack of real democracy with its impacts on public health policy and leadership; and thirdly the cultural inadequacy of the ethical principles of research and clinical work.

These three major challenges belong to the realm of the social determinants of health, far from the domain of clinical intervention. They heighten the need for social healing without which most African countries will not improve health indicators. Addressing health challenges locally requires knowledge of society and global institutions, as distinct from knowledge of diseases. The function of global health ethics in African countries should involve both a far more delicate and dynamic balancing act, requiring the identification of local values and public health policies through valid research and the provision of complementary support and guidance to promote justice, accountability, and preventive practices in an inherently unstable environment. A process that points to the scientific and interdisciplinary approach to policy and intervention which suggests, at the institutional level, inter-sectoral and multi-level approaches to intervention, should be devised.

GOVERNANCE OF HEALTH SYSTEMS AND LEADERSHIP

Just like other social institutions, healthcare systems are affected by negative public values and the lack of governments' political commitment to people's welfare. Good leadership opposes greed to promote an organizational culture that gives preeminence to the national interest. Program implementation often needs research for effective realization. Health interventions such as diagnostics, access to care, and public health programs are often not grounded in prior research. Similarly, prevention is not sustained by a well-thought out data collection and epidemiologic surveillance system. For example, implementation research to discern the socio-historical and cultural roots of the under-utilization of healthcare services, the lack of basic infrastructure, and the failures of previous public health interventions is rarely launched. Implementation research in poor or rural areas should have a social action component, so as to address both disease etiology and the social roots of implementation failure.

Good governance is critical to public health because it shapes the institutional and legal environment of healthcare systems. Governance determines sectoral and inter-sectoral dialogue, shapes the coordination of all activities by the ministry of health, and influences priority setting. The financial governance is often weakened by repeated suspensions of the financial execution of budgeted expenditure and by non-compliance with the legislation governing the management of public finances, which prevents a timely execution of programs. The insufficiency of resources allocated to the health sector is often due to poor planning and management of data at all levels, less involvement of regional health commissioners in the process of budgeting, insufficient coordination between the state and its financial partners, and the mobilization of aid on the basis of national programs and cooperation projects with their own management mechanisms.

The technical governance of the healthcare system is also poor. The sector lacks clear prospects supported by a well thought-out system of coordination by the ministry of health, providing a framework for donor investments and partners so as to avoid fragmentation of the health sector. Policy and leadership failures are determining factors for healthcare system fragmentation.

FRAGMENTED AND INEFFICIENT HEALTH SYSTEMS

Health sector fragmentation continues to impede health system strengthening in most African countries. The fragmentation results from, and has been exacerbated by, historical events, health policy choices, and disease response options. Many factors justify fragmentation: persistence of endemic diseases or occurrence of epidemics requiring vertical interventions; the impact of international lending conditionality and structural adjustment programs on health systems and outcomes; persistence of parallel systems managed by non-state actors;[3] high dependence on external funding to support the healthcare sector;[4] the effects of wars and civil strife on infrastructures and service delivery.[5]

I identify several areas of fragmentation.[6] The first area is that of health sector policy and planning. In most African countries, fragmentation across governance structures, funding, and external actor engagement continues to challenge the efficiency and coherence

---

[3] See Irene Akua Agyepong et al., "The Path to Longer and Healthier Lives for All Africans by 2030: The *Lancet* Commission on the Future of Health in Sub-Saharan Africa," *Lancet* 390, no. 10114 (2018): 2803–2859.
[4] See Alexander Kentikelenis et al., "The International Monetary Fund and the Ebola Outbreak," *Lancet Global Health* 3, no. 2 (2015): doi.org/10/1016/S2214-109X(14)70377-8.
[5] See Organization for Economic Cooperation and Development, "Monitoring the Principles for Good International Engagement in Fragile States and Situations Country Report 5: Sierra Leone," www.oecd.org/countries/sierraleone/44653693.pdf.
[6] See Arwen Barr et al., "Health Sector Fragmentation: Three Examples from Sierra Leone," *Globalization and Health* 15, no. 1 (2019): doi.org/10.1186/s12992-018-0447-5.

of health sector activities and impedes sustained health system strengthening. Moreover, fragmentation results from the lack of preparedness in front of public health challenges. Instead of well-thought-out policies and interventions prompted by data gotten from the epidemiologic surveillance system, African countries often react to urgent situations and needs. Rapid policy and planning development in the context of epidemics often increases health systems fragmentation. The recent Ebola crisis in Mano River countries and in the Butembo area in Democratic Republic of Congo follow the logic of quick policy and planning as a consequence of unpreparedness.[7] The same assumption can be made about the tri-disease programs under the supervision of the Global Fund.[8] Even though efforts to integrate this program into the healthcare sector have been advocated by local and international experts and actors, the need to capitalize on the expertise gained from this vertical program in most of Africa is plunged into chaos by the lack of resources and poor policing.

The second area of fragmentation is governance of partner engagement. Dependence on external development partners has been shown to jeopardize the sustainability of existing progress, undermine national and community ownership, and distort the national agenda in countries in sub-Saharan Africa.[9] In other sub-Saharan African contexts, it has been noted that systems' weaknesses, gaps in local capacity, and perceived lack of transparency have been found to contribute to the external actors' decision to implement parallel structures and vertical programs that introduce health sector fragmentation.[10]

The third area of fragmentation is concerned with key health system structures. As reported by other global health initiatives, key barriers to improving service delivery include weak drug and medical supply systems.[11] Donors' responses to this challenge often include establishing parallel supply chains to quickly meet the needs of their program. The already inadequate healthcare systems of African

---

[7] See World Health Organization, "Global Health Observatory Data Repository," apps.who.int/gho/data/view.main.

[8] The three diseases are AIDS, tuberculosis, and malaria. See Global Fund, "Accelerating the End of AIDS, Tuberculosis and Malaria as Epidemics," www.theglobalfund.org/en/.

[9] See Sarah Wood Pallas and Jennifer Prah Ruger, "Effects of Donor Proliferation in Development Aid for Health on Health Program Performance: A Conceptual Framework," *Social Science & Medicine* 175 (2017): 177–186, at 179.

[10] See John Adwok et al., "Fragmentation of Health Care Delivery Services in Africa: Responsible Roles of Financial Donors and Project Implementers," *Developing Countries Studies* 3, no. 5 (2013): 92–96, at 92.

[11] See Ricarda Windisch et al., "Scaling up Antiretroviral Therapy in Uganda: Using Supply Chain Management to Appraise Health Systems Strengthening," *Globalization and Health* 7 (2011): doi.org/10.1186/1744-8603-7-25.

countries, especially in sub-Saharan Africa, have been further damaged by the migration of their health professionals. Nowhere is the deficit of human resources for healthcare more severe than in sub-Saharan Africa, where the ratio of healthcare workers to the population is the lowest worldwide.[12] Forty-six out of the forty-seven countries within this region have significantly less than 2.28 physicians or nurses per one thousand people, which is widely regarded as the minimum threshold required to deliver basic health services.[13] This region of the world also carries nearly twenty-four percent of the world's disease burden while containing only three percent of its healthcare workforce and only one percent of its financial resources for healthcare.[14] The lack of political will to improve the health sector is perceptible. In 2001, in the declaration they adopted at the Abuja Summit in Nigeria (April 24–27, 2001), African leaders decided to devote at least fifteen percent of the annual budget to improving the healthcare sector. Ten years after the Abuja commitment, only Rwanda (18.5 percent), Botswana and Niger (17.8 percent), Malawi (17.1 percent), Zambia (16.4 percent), and Burkina Faso (15.8 percent) were able to honor their commitment. In the same period, four countries were clearly on track to meet the Abuja target: Swaziland (14.9 percent), Ethiopia (14.6 percent), Lesotho (14.6 percent), and Djibouti (14.2 percent).[15]

Vertical programming is another major cause of healthcare system fragmentation. With the persistence of endemic disease and the emergence of new epidemics, there has been an inflation of vertical interventions in many African countries. These initiatives have mobilized additional resources for health and boosted the action of community organizations and the private sector, but the creation of all these new entities as well as the presence of global health institutions complicates the task of governments on the ground. Vertical programs absorb some of the health workforce, creating a quantitative shortage of health professionals in public health services. This shortage increases with the demotivation of staff working in precarious conditions. Local professionals are often attracted by the working

---

[12] See World Health Organization, "World Health Report 2006: Working Together for Health," www.who.int/entity/whr/2006/whr06_en.pdf?ua=1.
[13] See World Health Organization, "World Health Report 2006: Working Together for Health."
[14] See Stella C. E. Anyangwe and Chipayeni Mtonga, "Inequities in the Global Health Workforce: The Greatest Impediment to Health in Sub-Saharan Africa," *International Journal of Environmental Research and Public Health* 4, no. 2 (2007): 93–100.
[15] See United Nations et al., "Note d'Information: 10 Ans Après 'l'Engagement d'Abuja' d'allouer 15% des Budgets Nationaux à la Santé," www.uneca.org/sites/default/files/uploaded-documents/CoM/cfm2011/com2011_informationnote10years-after-theabujacommitment_fr.pdf.

conditions of aid-financed programs to the detriment of the public system that stands out for its inefficiency due to a lack of coordination between stakeholders among themselves and with recipient governments.

Moreover, traditional donors are poorly coordinated. In many African countries, the influx of health non-governmental organizations (NGOs) reinforces the fragmentation of the national health system through activities that divert financial and human resources away from the public sector. NGOs often create structures that parallel public services. The implementation of the Paris Declaration on Aid Effectiveness can help protect health systems from fragmentation and inefficiency due to vertical interventions. This declaration challenges donors to support the scaling up of effective programs and projects that strengthen strategies aiming at developing operational health systems capable of assuming vertical interventions.[16] The implementation of sectoral strategies, co-financed by donors, addresses only partly these problems since aid is not always aligned with the priorities of recipient governments. Hence, vertical programming is not conducive to the integration of the health sector strategy into development strategies. The polarization of these strategies in the healthcare sector alone overshadows the interactions between health and its social determinants.

ETHICAL CHALLENGES, GLOBAL HEALTH, AND FOREIGN CULTURES

The presence of global actors or institutions in African countries raises important ethical issues related both to public health practice and research. In order to face these ethical challenges, ethics cannot be understood as a second-level field of inquiry. Ethical values framed within a global context and local cultures of host countries should not only be part of the curriculum of global health practitioners, but they should also influence policy and practices at their formative stage. With regard to health systems, "Ethics can be viewed and studied more broadly as an integral component of health systems development. In this form, ethics is an organizational, development-oriented force that provides both methodological and motivational support to public-health practitioners and policy-makers."[17]

---

[16] See Organization for Economic Co-operation and Development, "The Paris Declaration on Aid Effectiveness and the Accra Agenda for Action," www.oecd.org/development/effectiveness/34428351.pdf.

[17] Adnan A. Hyder et al., "Integrating Ethics, Health Policy and Health Systems in Low- and Middle-Income Countries: Case Studies from Malaysia and Pakistan," *Bulletin of the World Health Organization* 86, no. 8 (2008): 606–611, at 607.

Global Health Ethics and Clinical Work

Global health clinicians trained within the biomedical framework may be shocked by the way local clinical workers strive to strike the balance between the individual patient and their family or community. In most African countries, the influence of social, cultural, and environmental factors on health decision making cannot be undermined. We have to take these factors into account in order to fully appreciate the moral dilemmas and health challenges in settings where individualism does not prevail.

African traditions present a view of the human person that is essentially relational; it is within the social network that the individual lives and acts as a free person. Hence, imposing autonomy in its radical form can be seen as a form of epistemological and moral imperialism to countries where medical paternalism prevails on account of patient beneficence and shared-responsibility for health promotion or due to cultural politics that favors power holders. The social matrix that shapes people's understanding of health, personhood, and power cannot be ignored. In the context of research, even though the research team may have received permission of recognized authorities, informed consent can be seen as a process that starts with a good dialogue with the local chief to build trust and ensure that researchers will not exploit local populations; and then extend to local populations. Similarly, in clinical settings, having my grandmother or my brother in the consultation room may not be perceived as an intrusion into my privacy. Very often, taking someone to a doctor's appointment is rather a sign of solidarity. It is an indication that the sick person does not suffer alone. The same logic may be applied to truth telling.

Global Health Ethics and Biomedical Research

Given the substantial differences in individual exposure to health risks and the availability of health protective resources as well as differences in disease burden, mortality, and morbidity at the population level, it is clear that illness in underdeveloped countries can be better understood using a "social causation of illness" perspective. The principles of respect for autonomy, beneficence, and justice are all premised on biological individualism proper to the biomedical rationality. In the African context, individualism cannot ensure the cultural and social relevance of the ethical principles of biomedical research. These principles need to be responsive to the context of poverty and capitalist exploitation because these contextual factors can increase vulnerability and exploitation. Not only knowledge of risks and benefit but also socioeconomic factors may determine individual participation in clinical trials. Consent may be given but not freely because it is given out of a search for reward due to financial incentives. Hence, protection of research subjects cannot be achieved by securing the provision of information to enroll consenting subjects in research endeavors. A formal provision of consent by the research subjects can simply mask the misery that inhibits their ability to consent freely.

The concept of "benefit"—taken as what is due to research participants—is a universal construct that needs to be contextualized by taking into consideration local cultures, different levels of poverty, and disease burden in different resource poor countries. This concept certainly includes the notion of individual good, which is an improvement in individual health status. However, benefit cannot be confined to a reward given to individual participants. As a prism for understanding benefit, individualism does not account for the complex context that gives rise to substantial differences in disease risks and burden. Beyond the implicit or explicit recourse to individualism as a framework for understanding the principles of research ethics which goes beyond the need to reward individual participants, the concept of benefit may be redefined in terms of public and common good. Within this new frame, the concept of benefit can rather be understood in terms of increased access to health services, improved access of basic social goods, expanded access to education, and lobbying for a better social and health policy environment.

Furthermore, ethical guidelines for research tend to undermine an international context and a world order dominated by the exploitative tendency of neoliberalist capitalism, not only in socioeconomic and political matters but also in the domain of clinical research. The capitalist system—which determines resource distribution, exchange logics, and power control that favors wealthy countries—cannot be seen as alien to the global difference in health status and disease burden both between countries and various parts of the world. It is common knowledge that "The spread of health risks and diseases across the world, often linked with trade or attempted conquest, is not new to public health or international health."[18] The economic exploitation that prevails in the capitalist system shapes the global and local distribution of resources and diseases as well as the health risks and vulnerability of those who live on the margins of the global market. Biomedical research is then carried out in a world where subjects from resource-constrained countries can easily be exploited and even used for the sake of multinational corporations and well-off populations from developed countries. It can be said that the concept of "justice" contained in the Belmont Report[19] remains blind in front

---

[18] Jeffrey P. Koplan et al., "Towards a Common Definition of Global Health," *Lancet* 373, no. 9679 (2009): 1993–1995, at 1994.

[19] National Commission for the Protection of Human Subjects of Biomedical and Behavioral Research, "Belmont Report: Ethical Principles and Guidelines for the Protection of Human Subjects of Research," or.org/pdf/BelmontReport.pdf.

of the vulnerability of the underprivileged population of poor countries.

Four Contextualized Principles

Four important principles flow from this analysis. The first one includes respect for persons and goes beyond it to consider the sociocultural matrix within which persons frame their universe of meaning. This principle may be called "the principle of respect for person and for the alterity of their culture." The fundamental value that ought to shape one's approach to a foreign culture is nothing else than respect and openness to learn before attempting any action. Respect for persons is the overarching principle in biomedical ethics. In Africa, within communitarian settings, human dignity cannot be respected if its cultural groundings are not given their due place. In working with diverse groups of people, global health practitioners should be aware that notions of health and disease differ across cultures and that they must understand those differences to do effective cross-cultural work. However, adopting these values is challenging when a health practitioner is faced with the task of containing an epidemic in a foreign context. Culture competence, acquired through training and familiarity with a culture and sustained by the respect due to foreign cultures, may help face such a challenge.

Awareness of foreign cultures ought to be an essential component of the educational preparedness of global health workers, and the inclusion of cultural sensitivity and awareness into the curriculum will promote competency within the global health realm. Being aware of a foreign culture does not mean that one knows everything or that one is fully competent. Cultural humility is based on the idea of focusing on self-reflection and lifelong learning. Cultural humility involves an ongoing process of self-exploration and self-critique combined with a willingness to learn from others. It means entering into relationship with other persons with the intention of honoring their beliefs, customs, and values. Cultural humility incorporates a lifelong commitment to self-evaluation and self-critique, to redress the power imbalances in the patient-physician dynamic, and to develop mutually beneficial and non-paternalistic clinical and advocacy partnerships with communities on behalf of individuals and defined populations.[20] This notion leads us to think about our preconceived notions of other people.

Global health has the potential of having a positive impact on important unmet needs around the world. However, I am concerned

---

[20] See Melanie Tervalon and Jann Murray-Garcia, "Cultural Humility Versus Cultural Competence: A Critical Distinction in Defining Physician Training Outcomes in Multicultural Education," *Journal of Health Care for the Poor and Underserved* 9, no. 2 (1998): 117–125, at 118.

about whether such initiatives could be unknowingly ineffectual or even create more problems than solutions. The risk of harm is particularly high when global health involvement is shrouded in a veil of naïve altruism or domineering heroism, perpetuating colonial dynamics whereby Western knowledge and viewpoints subjugate alternatives. To maximize the potential for positive global health involvement, while heightening awareness of the potential for ineffectiveness or even harm, global health workers should engage in reflective practice premised on the concept of cultural humility.[21]

The lack of basic respect for a culture can be counterproductive, as was seen during the recent Ebola epidemic in Mano River countries. The intervention teams in Guinea's Forest Region simply ignored the local culture and the ontological regime from which it emerges. Without accentuating this conflict, I want to stress that the human person cannot be reduced to his or her biological life. For people living in the Guinea's Forest Region, "A dead person can remain a person to be cherished because she remains a person endowed with a capacity for action on the world and for the fate of those who survive her."[22] Notwithstanding this anthropological argument, a more political hermeneutics of the crisis could have been developed so as to uncover the social elements that patterned the crisis. The treatment of bodies imposed by the authorities was equated to the little consideration that the elites have for people. Thus, the initial response increased mistrust which prevented people from seeking help in government-run institutions and reduced the level of community contribution to relief efforts. Relief workers had to learn once again that effective intervention needed to be culturally acceptable. Interventionists thought that they could change long-standing and deeply-engrained burial and funeral practices, which include washing of the body after death. These practices, which involve close contact with infected corpses, contributed to the spread of the disease. Quarantine, social distancing, mandatory cremation of bodies, and other top-down solutions were used to avert incident cases arising from these practices instead of using credible individuals from the communities to reinforce health promotion messages that were tailored to local cultures.[23] The relief workers not only lacked the basic set of abilities,

---

[21] See Shaun R. Cleaver et al., "Cultural Humility: A Way of Thinking to Inform Practice Globally," *Physiotherapy Canada* 68, no. 1 (2016): 1–4, at 1.

[22] Frédéric Le Marcis, "'Traiter les Corps Comme des Fagots': Production Sociale de l'Indifférence en Contexte Ebola (Guinée)," *Anthropologie & Santé* 11 (2015): doi.org/10.4000/anthropologiesante.1907. The translation is mine.

[23] See Sylvain Landry Faye, "L''Exceptionnalité' d'Ebola et les 'Réticences' Populaires en Guinée-Conakry: Réflexions à Partir d'une Approche d'Anthropologie Symétrique," *Anthropologie & Santé* 11 (2015): doi.org/10.4000/anthropologiesante.1796.

skills, behaviors, and policies that enables practitioners to work effectively in cross-cultural situations but also cultural humility.

The second principle—the "principle of social justice"—places the quest for good health within the context of exploitative neoliberal systems, local cultural politics, and national structural injustices. Far from being a context-blind principle, it aims at pointing out criteria for equitable chances for being healthy or acceptable treatment of research subjects. It takes into account local, national, and global forces that shape the course of life in African countries. Even though what is due to the individual is not left out, this principle goes beyond distributive justice in clinical settings and research sites to conceptualize public health challenges in relation to social policy and global health initiatives so as to confront systemic injustices. This principle is directly related to the principle of "public benefit" in that 'benefit' is framed within the larger context of local and global exploitation.

The third principle can be called "principle of public benefits." This principle is framed within a sociological approach to health to emphasize that risks, benefits, and equity cannot be defined only in terms of individual health but also must be done in relation to contextual factors. The sociological model is premised on the non-separation of nature (i.e., disease) from society.[24] Mark Tausig et al. emphasize that "sociological explanations of the causes of illness give priority to inequalities in social structures based on economic status (including poverty), as well as gender and race/ethnicity thereby linking nature and society."[25] The principle of public benefits is a political and socioeconomic critique of the dominant understanding of "justice," in research and biomedical ethics, given well-established patterns of exploitation and oppression of the underprivileged. Such a principle challenges the individualistic understanding and account of benefits in places where social and global inequalities not only determine access to care but eminently structure differences in disease burden. Consequently, a thicker approach to benefit brings to the forefront the researchers' moral agency as well as the social responsibility of research institutions.

The fourth principle focuses on "local capacity building." Premised on a contextual understanding of differences in disease risks and burden and on the one-sidedness of the individualistic understanding of benefit, this principle states that building capacity to promote healthcare sustainability and health system development in resource stressed countries will have a lasting impact on individuals' participation and on local community at large. This principle stresses

---

[24] See Mark Tausig et al., "The Bioethics of Medical Research in Very Poor Countries," *Health* 11, no. 2 (2007). 145–161, at 146.
[25] Tausig et al., "The Bioethics of Medical Research," 146.

that it is unjust to compensate individual participants with a little money. Such an amount cannot be given to research participants in the country where the research institution normally operates. However, with their money, institutions can address some of the consequences of structural injustices that increase the risks for infection or reduce access to care. Local capacity building may entail improving human capital, reinforcing existing health infrastructures, and encouraging economic self-reliance through income generating opportunities.

CONCLUSION

Global health intervention presents the prospect of improving public health in most African countries faced with a fragmented and inefficient healthcare system. The fragmentation of the health sector is due to historical and unforeseeable reasons. Vertical programming has been perceived as a key solution to face some urgent public health challenges. However, good leadership, adequate policy, and innovative solutions sustained by coordinated external interventions, and implemented with context-based ethical principles, may yield good fruits. In addition, in order to respond to health challenges, ethical guidelines for research and biomedical intervention ought to be responsive to local cultures and socioeconomic problems faced by local populations. M

Jacquineau Azetsop, SJ, holds a PhD in social ethics from Boston College with a thesis focusing on health inequality and social justice. He also earned a Masters of Public Health from the Bloomberg School of Public Health at Johns Hopkins University and a Masters in theological ethics from the Weston Jesuit School of Theology. Jacquineau taught health policy, social aspects in public health, medical deontology, and bioethics at the N'Djamena University's School of Health Sciences in Chad and at the school of medicine of the Catholic University of Mozambique in Beira. Since 2014, he has been teaching courses on socioeconomic inequality, structural interventions in public health, social suffering, and the sociology of health and illness at the Gregorian University in Rome, where he is the current dean of the Faculty of Social Sciences. His broad research interests include cultural practices and human rights as related to health promotion, social ethics and health systems challenges, AIDS and social justice, Christianity and politics in Africa. He has published widely, including more than twenty research papers and book chapters and two books: *Structural Violence, Population Health and Health Equity* and the edited volume *HIV and AIDS in Africa: Christian reflections, Public Health and Social Transformation.*

Public Health Concerns in India: A Review

Stanislaus Alla, SJ

AS A NATION, DIVERSITY AND PLURALITY define India, and its complexity is shared by several Asian nations. Being colonized territories, these countries developed similar administrative structures and systems, and they also embrace an overlapping spectrum of cultural views, challenges, and response-mechanisms. In this chapter, I limit myself to India, referring to the state of peoples' health, public discourse on health concerns, governmental policies and interventions, and their effectiveness. Cultural components continue to play a critical role, and they deserve a special attention. Apart from poverty and population concerns, illiteracy and ignorance, superstition and corruption are still rampant, and they severely limit the efforts to make healthcare accessible to large sections of the population. As in other places in the world, climate change and pollution are having a devastating impact on the lives of the poor in India. Here, first, I briefly describe the state of people's health as a story of success; second, I point to what ails healthcare services in India; and, finally, I suggest a few ways for enriching a public discourse on health.

A SUCCESS STORY—AT LEAST PARTIALLY

To begin on a positive note, it is important to recognize that in spite of many shortcomings, multiple indices reveal that healthcare has been made accessible to several sections of the Indian population. Mortality rates of mother and child have come down, and the average life expectancy has improved from thirty-two years in 1951 to sixty-eight years in 2016.[1] Successive governments made provisions to supplement nutrition needs to the poor, especially the expecting and lactating mothers and the malnourished children.[2] Being largely

---

[1] See S. Mahendra Dev, "Social Sector in the 2019 Union Budget," *Economic and Political Weekly* 54, no. 33 (2019): 43–44. Noteworthy, however, are several regional disparities in the life expectancy. For instance, in Kerala, it is seventy-five years; in Uttar Pradesh, it is sixty-five years; and in Madhya Pradesh, it is forty-five years.
[2] See T. K. Rajalakshmi, "Cosmetic Measures," *Frontline*, www.frontline.thehindu.com/the-nation/cosmetic-measures/article10107321.ece.

responsible for the social sectors since India's independence, various States and Central governments have been providing funds to establish and maintain hospitals both in rural and urban areas in order to make health care services available to many. At these government hospitals, irrespective of their financial status, people are offered diagnostic services, medicines, and when possible, different kinds of surgeries are also performed. In the past, the government hospitals used to do well, and now they still strive to do so in a very limited way. In the Indian administrative structure, Central and State governments share responsibility for healthcare (as well as education), and where the States have strong control mechanisms, the results have been remarkable.[3] Kerala tops in the country, scoring high on all health parameters, while other states like Andhra Pradesh, Maharashtra, Goa, and Delhi have impressive records.

An innovative flagship health care program titled *Rajiv Arogyasri* was initiated in 2007 in the state of Andhra Pradesh by Y. S. Rajasekhar Reddy, the then Chief Minister of the state.[4] The primary objective of the program is: "to provide free quality hospital care and equity of access to BPL [below poverty line] families by purchase of quality medical services from identified network of health care providers."[5] Chiefly, it is aimed at providing financial security against catastrophic health expenditures to the rural and urban poor.

Eventually recognizing its usefulness and popularity, some other States have imitated this program and, with some variations, the Prime Minister of India, Narendra Modi, announced a similar program in 2018. This *National Health Protection Scheme* is also variously called as *Ayushman Bharat* or *PMJAY*.[6] The Prime Minister's program evolved from the earlier health mission programs and subsumes various health insurance policies including those of senior citizens and employees of the State and Central governments. Ambitiously, it plans to cover a hundred million poor and vulnerable families (i.e., about five hundred million people, presumably making it the largest health care program in world) promising to provide up to five lakh rupees

---

[3] See Editorial, "Scoring on Health: On Health Index 2019," *The Hindu*, www.thehindu.com/opinion/editorial/scoring-on-health/article28159373.ece.
According to this editorial, the NITI Ayog Index, operated by the central government, includes parameters such as "neonatal and infant mortality rates, fertility rate, low birth weight, immunization coverage and progress in treating tuberculosis and HIV" to assess the health and wellbeing of the people.

[4] Details of the program are found at the official YSR Arogyasri Health Care Trust website: www.ysraarogyasri.ap.gov.in.

[5] The objectives are found at the program's official website, cited above.

[6] At the official website one can find several details. See India Government, "Ayushman Bharat," www.india.gov.in/. For a critical analysis of the program, see T. K. Rajalakshmi, "Health Care Hoax," *Frontline*, www.frontline.thehindu.com/cover-story/health-care-hoax/article10074189.ece.

(about $7,000) per family annually. While funding and implementation mechanisms are true challenges, the program itself claims to be game changer.

Many NGOs and the Catholic church[7] (as well as the other churches and service wings of other religions) collaborate with the government in providing healthcare services to the needy as well as in realizing special targets. In some instances, where the purpose is to contain and eradicate a particular illness, the mission has been largely successful. Through immunization programs, polio is almost eradicated across the country and malaria and other illnesses are largely contained, thanks to the efforts of the government and its collaborators. Similarly, many acknowledge and appreciate the efforts of the government and its partners in bringing awareness about HIV/AIDS. Declaring it as a national emergency project, the government intervened and made an all-out effort to spread awareness about HIV, and the various branches of media were employed successfully. Proper allocation of funds, sustained campaigns, and commitment of the healthcare personnel have saved millions of lives through these innovative programs.[8]

While the nation's healthcare sector is heavily privatized (nearly eighty percent have to pay out of pocket), India boasts of having very good super-specialty hospitals that offer world-class facilities.[9] In the last two decades, one can note the expansion and growth of this sector in all major cities. The wealthy Indians, and many from countries far and near, are able to receive high quality healthcare services. Similarly, pharmaceutical and biotech firms carry out research in many cities and manufacture and supply medicines globally. Hyderabad, along with other cities, is a major global hub for this business.

WHAT AILS HEALTHCARE SERVICES IN INDIA?

If there are several areas where one notices significant improvements across the country that enable people to have access to health care, there are other areas that need urgent attention. Here I point to three concerns: a) budgetary allocations, b) food and water contamination, and c) diversion tactics. More funding for public health is essential if the nation hopes to have healthy people who can work for the prosperity of the country. For various reasons, budgetary allocations have been on the decline, and the governments ought to

---

[7] For details, see Bishops' Conference of India Commission of Health, *Sharing the Fullness of Life: Health Policy of the Catholic Church in India* (New Delhi: CBCI Centre, 2005).
[8] For details, see the government's site on National AIDS Control Organisation, www.naco.gov.in/.
[9] See R. Ramachandran, "Retreat of the State," *Frontline*, frontline.thehindu.com.

step up. More than most countries in the world, food and water pollution levels are dangerously high in India and, unless drastic measures are taken, soon the country will be at a point of no return. In this regard, it is indispensable to change governmental policies as well as the people's cultural mindset. A risky and dangerous trend, promoted by the Hindutva forces, diverts the people's attention. For instance, instead of assigning funds for public health concerns, the government allocates money for the welfare of cows and, instead of paying attention to starving children, it punishes the journalist who brought the issue to light. To counter such callous negligence, public awareness and civil movements are needed to make governments accountable and responsible.

Budgetary Allocations

Compared to several developed and developing nations (some of them allocate nearly five percent of the nations' budgets for health care needs), the Indian government allocates disappointingly less for the social sector, which includes healthcare. The States combined allocated only 1.5 percent of GDP in 2018 – 2019 (1.2 percent in 2013-2014) for the whole health sector that incorporates medical and public health, family welfare, water supply, and sanitation. The government hopes to increase its allocation to two percent while the WHO recommended ideal is five percent. Since the 1990s a big push to privatize the health sector (education as well) made health services less accessible and less affordable to the needy: along with it, the poor quality of health care delivery systems makes India's health indicators lag behind in comparison to the BRICS (i.e., Brazil, Russia, India, China, and South Africa) partners[10] and, in some instances, South Asian neighbors fare better in this regard. Significant regional, social, and gender disparities exist in the country, making access to healthcare even more difficult.

Compared to the required funding for the ambitious Prime Minister's health insurance coverage (PMJAY), the money allocated (Rs 6,000 crore–about 794 million USD) is trivial. Similarly, the promised financial assistance to the needy mothers and children is too little or it reaches them too late.[11] Jean Dreze states that "massive increase in public health expenditure and a radical revamp of primary health infrastructure are needed for moving towards universal health coverage."[12] This engagement is imperative in a country where inequality is increasing. Oxfam Inequality Report 2019 observes the existence of gross inequalities in India and states that while "the top 1 percent own 51.53 percent of national wealth, the bottom 60 percent

---

[10] See Dev, "Social Sector in the 2019 Union Budget," 43.
[11] See Dev, "Social Sector in the 2019 Union Budget," 44.
[12] Dev, "Social Sector in the 2019 Union Budget," 44.

own merely 4.8 percent."[13] The observation Prabhat Jha made in 2014 is still relevant: "India already spends about 6 percent of its GDP on health care. But 80 percent of this is out-of-pocket and drives over 40 million Indians below poverty line every year."[14] Since the privatization of the health sector drives more people into poverty, ways have to be imagined to help them through public funding in order to support their health and their families.

Food and Water Contamination

The use of fertilizers, chemicals, and pesticides that began in 1970s and 1980s and expanded unabated across the country, covering every sector in agriculture, promised great harvests and profits to the farmers. They replaced natural fertilizers and pesticides, largely in use until then. The agricultural transformation ushered in what is popularly known as the Green Revolution, and the increased productivity brought more food to the plates of millions. In its train came the excessive use of water and of fertilizers and chemicals. Over the ensuing decades chemical residues filtered into the soil and into the bodies of water across the country. Partly due to these developments, there is an undeclared health emergency in the country as lifespans began to decline rapidly. Citing a recent study, *The Times of India* declared "food and water contamination as one of the biggest health challenges facing the country."[15] However, the study noted that even though "food and water-borne diseases extract a heavy price on the Indian population and the country's economy," people do not recognize it because in the popular imagination "spurious food is not adequately captured as a risk factor, partly because food-borne diseases go largely undetected and unreported."[16] The study demonstrated that, in 2016 – 2017, 4.8 percent of GDP is the price India paid to treat food and waterborne diseases. Over all, including the worktime lost, these diseases cost India 7.3 lakh crore rupees (about one million USD).

Besides communicable diseases, in the recent times one non-communicable disease that got greater attention is Chronic Kidney Disease (CKD), which causes a large number of renal failures and

---

[13] Akhil Kumar, "Nine Richest Indians Now Own Wealth Equivalent to Bottom 50% of the Country," *The Wire*, www.thewire.in/rights/9-richest-indians-now-own-wealth-equivalent-to-bottom-50-of-the-population.
[14] Prabhat Jha, "Quality Health Care with Public Funds," *The Hindu*, www.thehindu.com/opinion/op-ed/quality-health-care-with-public-funds/article6101801.ece?homepage=true.
[15] Karthic C. Iyer, "How Spurious Food, Unsafe Water Cost India 7l Crore," *The Times of India*, www.pressreader.com/india/the-times-of-india-mumbai-edition/20190828/282995401535640.
[16] Iyer, "How Spurious Food, Unsafe Water Cost India 7l Crore."

deaths.[17] To substantiate the findings, Maya quotes from an editorial published in the *New England Journal of Medicine*:

> What we do know for certain is that CKDu is related to heat exposure and dehydration, although exposure to agrochemicals, heavy metals, and infectious agents, as well as genetic factors and risk factors related to poverty, malnutrition, and other social determinants of health may also contribute.[18]

Environmental degradation is noted to be a major cause for renal failures. By quoting a nephrologist in Kerala, Maya reports that, "From 30-40 patients a day way back in 2006, today I see around 300 or 400 patients in my OP. What is worrying is the steady increase in this group of patients, who constitute 20-30 percent of the OP cases."[19] The increased occurrence of kidney failures in people in their 30s and 40s underscores that India is grappling with another epidemic.

Related to the food and water contamination is air pollution, which began to get global attention in recent times. Alarmingly, Delhi is declared as the most polluted city in the world, and India is home to fifteen out of the twenty most heavily polluted cities across the world.[20] The same report suggests that Delhi's "toxic air is caused by vehicle and industrial emissions, dust from building sites, smoke from the burning of rubbish and crop residue in nearby fields." Air, water, and food pollution lie at the root of non-communicable diseases like cardiovascular diseases and diabetes. Sadly, these diseases are increasing, with a heavy toll on the poor.

Diversion Tactics

A concern that received insufficient attention is the lack of political will in maintaining and expanding government health centers. Influenced by the right-wing Hindutva forces, several governmental bodies are not able to implement decisions that are informed by scientific data and that aim at being objective and transparent. As

---

[17] See C. Maya, "The New Reality Called Climate Medicine," *The Hindu*, www.thehindu.com/news/national/kerala/the-new-reality-called-climate-medicine/article29326418.ece.

[18] Maya, "The New Reality Called Climate Medicine." The author quotes Cecilia Sorensen, and Ramon Garcia-Trabanino, "A New Era of Climate Medicine—Addressing Heat-Triggered Renal Disease," *New England Journal of Medicine* 381, no. 8 (2019): 693–696, at 693. The acronym CDKu means "chronic kidney disease of unknown origin."

[19] Maya, "The New Reality Called Climate Medicine." The acronym OP means "outpatient practice."

[20] The data are available at: Mayank Bhardwaj and Sanjeev Miglani, "New Delhi Is World's Most Polluted Capital, Beijing Eighth," *Reuters*, www.reuters.com/article/us-india-pollution/new-delhi-is-worlds-most-polluted-capital-beijing-eighth-idINKCN1QM1FH.

noted, if the allocation of funds is a challenge, an equally important factor is the lack of clarity on where the budgeted funds would be directed to. Some governments, like the state of Rajasthan, went to the extent of having a Ministry for Cows. Funds have been assigned to study the properties of cow urine and breed more cows that have "healing" properties. Diverting money from health budgets for such projects illustrates the scant respect some policy makers have for scientific approaches. While funding such studies is not unhelpful, what is worrisome is the fact that some urgent concerns—such as caring for ill and malnourished children—will go underfunded.

As already indicated, the current political disposition seems to be in a "cover up and deny" mode. The Modi government came to power with the slogans of "minimum government" and "maximum governance," but concretely it meant that government will reduce funding for social sectors and welcome private enterprises to develop healthcare facilities. Moreover, the government opposes any criticism of its programs and policies, and if anyone exposes structural or administrative defects, not only is the issue ignored but those reporting will likely be punished. Most shocking incidents were the arrests of, first, a doctor who broke the story about how inadequate facilities led to the death of several infants at a government hospital[21] and, second, a news reporter who exposed the poor quality of food served to undernourished children.[22] When the poor fall sick, not infrequently they are blamed for their carelessness and negligence rather than finding the administrative personnel accountable and responsible for lack of care provided to them. Socio cultural factors such as ignorance, illiteracy, and superstition add fuel to the fire. Discrimination based on caste, creed, gender, region, and political affiliation deprives and denies these designated "others" of needed healthcare services.

Additionally, even though the "universal health insurance schemes" of both States and the Central governments are to be lauded, studies show that they have largely catered to the middle classes than to the actual rural poor.[23] One of the reasons why the governments do not antagonize the middle classes is possibly because these classes tend to construct and influence public political discourses. The poor are less likely to question the government's practices, taking illnesses

---

[21] See Shone Satheesh, "Gorakhpur Doctor Jailed for Infant Deaths Claims Many Adults May Have Also Died When Oxygen Ran Out," *Scroll.in*, scroll.in/pulse/883944/gorakhpur-doctor-jailed-for-infant-deaths-claims-many-adults-may-have-also-died-when-oxygen-ran-out.

[22] See Press Trust of India, "Classic Case of Shooting Messenger: Editors Guild on Journalist Booked for 'Salt-Roti' Report," *India Today*, www.indiatoday.in/india/story/classic-case-of-shooting-messenger-editors-guild-on-journalist-booked-for-salt-roti-report-1594650-2019-09 02.

[23] See Rajalakshmi, "Health Care Hoax."

as their fate. Moreover, instead of developing governmental facilities where people can be treated, the government selects private health facilities to be the major service providers, breeding corruption. Similarly, studies show that, in implementing the insurance schemes, the government is spending more on secondary and tertiary care (where, again, the middle class will profit most) and not on primary health care (where the poor are likely to benefit more). Hence, through the insurance schemes, taxpayers' money is handed over to the private sector and more is spent on "curative" than "preventive" aspects of healthcare.[24]

ENRICHING PUBLIC DISCOURSE ON HEALTH

We briefly noted successes and limitations of healthcare services in India. One of the factors that is understudied in India is how the notions of dignity, equality, and human rights are connected to healthcare. Here I highlight two points. First, by noticing that there is a conflict between constitutional and cultural values in India, public discourse should discuss, examine, and address such a conflict. Second, human rights are not simply a legal concept. They should also be considered as a moral compass, which could help in promoting health by aiming at providing what is required to foster health and by prohibiting what is harmful for people's health. In such a way, human rights are essential for advancing the well-being of all.

Some scholars argue that there is a difference between constitutional values, which declare that all Indians are endowed with human dignity and equality, and cultural values, which are founded on the notion of caste and other related religious and cultural worldviews stressing that we are not born with universal and inalienable dignity, equality, and rights.[25] This is more than a theoretical debate as hierarchical and caste mentalities dominate and continue to deprive the Dalits, tribals, and other marginalized sectors of dignity, equality, and rights. This faulty understanding of the self and of the other (considering oneself to be superior to all, while others are inferior) is

---

[24] See Rajalakshmi, "Health Care Hoax."
[25] See Prabhat Patnaik, "From Revolution to Counter Revolution," *Frontline*, frontline.thehindu.com; Venkitesh Ramakrishnan, "Creed above Country: Rise of the Right," *Frontline*, frontline.thehindu.com. In the commemorative volume of *Frontline*, celebrating the 70th year of India's Independence, these essays discuss several substantive issues that shape the nation. For a Hindu ethical worldview, see S. Cromwell Crawford, *The Evolution of Hindu Ethical Ideals* (Honolulu: The University of Hawaii Press, 1982); Manu, *The Law Code of Manu: A New Translation Based on the Critical Edition*, trans. Patrick Olivelle (Oxford: Oxford University Press, 2004). To articulate Indian life and values, some members of the Constituent Assembly—mandated to draft the Constitution of India—preferred the Law of Manu—which upholds and endorses hierarchy and inequality—to the secular Constitution.

partially responsible for our collective failure in offering appropriate healthcare services to all. Indians need to resolve their social inequalities and remind themselves that the principles and values enshrined in the Constitution are universal, non-negotiable, and binding. The Preamble of the Indian Constitution strives for

> JUSTICE, social, economic and political;
> LIBERTY of thought, expression, belief, faith and worship;
> EQUALITY of status and of opportunity;
> and to promote among them all FRATERNITY assuring the dignity of the individual and the unity and integrity of the Nation ....[26]

On the contrary, the notion that karma or fate decides who is to be well or ill and believing that both others and structures are not responsible for one's state of health needs urgent correction. All ought to recognize both the individual and collective responsibility and accountability, especially of those in administration. Only then will people be in a position to confront, question, and receive what is due to them. Upholding and discussing constitutional values and examining health concerns and challenges in that light are critical challenges in the Indian context. As a result, people could be empowered with a sense of dignity and equality, and their moral agency could be strengthened.

Moreover, in Hinduism, human rights are considered problematic,[27] and this has negative consequences for promoting people's health. Because dignity and equality are inseparable from fostering health, implementing justice and protecting human rights are integral to advancing public discourse on health and specific policies. Discrimination and denying access to healthcare services should be questioned on human rights grounds. Discussions and debates on people's right to pure air, clean water, and healthy and nourishing food are critical. The government and its various structures and systems are required to strive for these goods by proactively and mandatorily doing what could be done, like acting against those who pollute air and contaminate water, soil, air, and foods.

CONCLUSION

Health is a complex good, and India faces many challenges to promote health for its citizens and the whole country. In this scenario, individual efforts, as well as those of people in authority, are important and essential. Health parameters are alarming and collective efforts

---

[26] India Government, "Constitution of India," *india.gov.in*, www.india.gov.in/sites/upload_files/npi/files/coi_part_full.pdf.
[27] See Satguru Sivaya Subramuniyaswami, "Hinduism's Human Rights Dilemma," *Hinduism Today*, www.hinduismtoday.com/modules/smartsection/item.php?itemid=3369.

are lacking. At the governmental level, the national vision and mission reflect the willingness to place checks and balances, and ensure legal provisions, but there is a lot that needs to be done on the cultural front. To confront environmental degradation, multi-cultural and multi-religious approaches are urgent and unavoidable. To renew and generate a vibrant discussion on public health needs and strategies is a priority. The Indian Constitutional values are a common ground, and they can potentially bring people of various persuasions together working out needed strategies to offer necessary healthcare services to every citizen, especially the needy. ■

Stanislaus Alla, SJ, is a Jesuit from Andhra Pradesh, India, and teaches at the Vidyajyoti College of Theology, Delhi. He obtained a licentiate in Moral Theology from the Alfonsiana Academy in Rome, and a PhD in Theological Ethics from Boston College. At Vidyajyoti he teaches courses in fundamental moral theology, virtue ethics, bioethics, and sexual ethics. He published on health care issues in the Indian context by engaging Hindu bioethics.

Journal of Moral Theology

A European Take on Global Public Health: Applying the Catholic Principle of Subsidiarity to Global Health Governance

Thana Cristina de Campos

2019 COVID, 2016 ZIKA, 2014 EBOLA, and 2009 H1N1 outbreaks have repeatedly shown that global public health needs a better (more effective and more efficient) governance approach to tackle Public Health Emergencies of International Concern (PHEIC). The World Health Organization (WHO), perceived as the global public health leader *par excellence*, is typically accused of not doing enough.[1] A proper leader, so it is claimed, should centralize its authority and powers to reverse the spread of these PHEIC.[2] I call this "the centralization approach." I argue against it. Centralization is not always the most reasonable, effective, and efficient form of leadership in global governance. To explore and argue for the decentralization approach to global public health, I introduce a traditional idea of the European Union (EU) governance: the principle of subsidiarity.

As a principle of Catholic social teaching, subsidiarity has been emphasized as an effective governance tool over centuries by various papal encyclicals. It establishes that, where families, neighborhoods, and local communities can effectively address their own problems, they should do so. When they cannot, then governments and other higher-level structures of power and authority should intervene and provide aid (i.e., *subsidium*). The term "sub-sid-iary"—which literally

---

[1] See Lawrence O. Gostin, "Ebola: Towards an International Health Systems Fund," *Lancet* 384, no. 9951 (2014): doi.org/10.1016/S0140-6736(14)61345-3; Lawrence O. Gostin and Eric A. Friedman, "Ebola: A Crisis in Global Health Leadership," *Lancet* 384, no. 9951 (2014): doi.org/10.1016/S0140-6736(14)61791-8; Ilona Kickbusch and Krishna S. Reddy, "Global Health Governance: The Next Political Revolution," *Public Health* 129, no. 7 (2015): 838–842; Tim K. Mackey, "The Ebola Outbreak: Catalyzing a 'Shift' in Global Health Governance?," *BMC Infectious Diseases* 16 (2016): doi.org/10.1186/s12879-016-2016-y; Lawrence O. Gostin et al., "The Normative Authority of the World Health Organization," *Public Health* 129, no. 7 (2015): 854–863, at 857.

[2] See Mackey, "The Ebola Outbreak"; Gostin et al., "The Normative Authority," 857.

means to "seat" ("sid") an activity down ("sub") as close to the problem as possible[3]—recognizes the value of first trying to solve social problems locally and moving up to higher levels of governance only as necessary. This Catholic principle was adopted as part of the EU law in the 1992 Treaty of Maastricht in order to prevent excessive centralization within the European system of governance. Its legal requirements are most clearly defined in Article 5(3) of the 1992 Maastricht Treaty, which states that "in areas which do not fall within its exclusive competence, the Union shall act only if and in so far as the objectives of the proposed action cannot be sufficiently achieved by the Member States, either at central level or at regional and local level, but can rather, by reason of the scale or effects of the proposed action, be better achieved at Union level." In short, the 1992 Treaty of Maastricht embedded EU law with the principle of subsidiarity to prevent excessive centralization in Brussels, to recognize the value of a plurality of social groups, and to require "higher (larger) groups to aid lower (smaller) groups, rather than to obliterate or subsume them."[4]

The idea conveyed by the principle of subsidiarity, however, has not been applied to other structures of global governance beyond the EU, but subsidiarity and its decentralization approach could be a particularly promising principle for global health governance. In this chapter, I analyze some of main features of the centralization approach—the mainstream position among most global health governance scholars—according to which the WHO should centralize more power and authority to be able to exercise better leadership in global public health. To challenge this approach, I introduce the idea of subsidiarity as a structural principle of governance and discuss its application to global health governance. In doing so, I explain why subsidiarity proposes a better (i.e., more reasonable, effective, and efficient) alternative than centralization. I conclude by affirming that the principle of subsidiarity justifies certain limitations to the roles and functions of the WHO and of other higher-level global health authorities by including the participation of local communities and, thereby, strengthening the overall framework for preparedness and response to PHEIC.

THE CENTRALIZATION APPROACH

Global health governance is the worldwide institutional structure whose main purpose is to orchestrate a wide range of independent stakeholders, comprising state and non-state actors, with different

---

[3] See Robert K. Vicher, "Subsidiarity as a Principle of Governance: Beyond Devolution," *Indiana Law Review* 35 (2001): 103–142, at 103.
[4] Nicholas W. Barber and Richard Ekins, "Situating Subsidiarity," *American Journal of Jurisprudence* 61, no. 1 (2016): 5–12, at 5.

capacities and self-regulating mandates.[5] Examples of global health stakeholders include: the World Bank; United Nations agencies (such as the WHO and the International Monetary Fund); civil society organizations (like Global Fund, Médicins Sans Frontières, Red Cross, Bill and Melinda Gates Foundation, etc.); pharmaceutical transnational corporations; global health donors; governments in developed, developing, and underdeveloped countries; the local communities affected by the global public health problem under consideration; and the hospitals, healthcare professionals, medical researchers, and even the media reporting on global health threats. These stakeholders are independent because the laws, policies, and programs they enact typically address one singular aspect of a multifaceted global health problem in a self-determining way.[6] Yet, they are willing to cooperate in addressing together global public health threats, like the recent PHEICs.

Although global health governance should serve the purpose of orchestrating this wide range of stakeholders towards an adequate global public health response that safeguards the global common good, lack of coordination among stakeholders is a fundamental defect of this governance structure. Lack of coordination leads to inadequate, delayed, and fragmented responses, as well as duplication of global public health services, waste, and unnecessary competition among stakeholders that could be working together in more efficient and complementary ways.[7]

While global health governance's lack of coordination among global health stakeholders leads to efficiency problems, its lack of inclusion and participation of local communities further leads to serious effectiveness problems.[8] Local communities, directly affected by the global public health problem under consideration, know better than outsiders what their specific health needs and most urgent concerns are. Not including affected communities in the decision-making process is therefore a sure way to respond less effectively to the problem. It is argued, however, that the chief cause of such exclusion is the donor-centered model of international development

---

[5] See Sophie Harman, *Global Health Governance* (Abingdon, UK: Routledge, 2012).
[6] See Jennifer Prah Ruger, "Global Health Governance as Shared Health Governance," *Journal of Epidemiology and Community Health* 66, no. 7 (2012): 653–661, at 653.
[7] See Ruger, "Global Health Governance," 654; Mackey, "The Ebola Outbreak"; Lawrence O. Gostin and Emily A. Mok, "Innovative Solutions to Closing the Health Gap between Rich and Poor: A Special Symposium on Global Health Governance," *Journal of Law, Medicine & Ethics* 38, no. 3 (2010): 451–458, at 451 and 453; Gostin et al., "The Normative Authority," 854, 857; Gostin and Friedman, "Ebola: A Crisis in Global Health Leadership," 1323; Kickbusch, and Reddy, "Global Health Governance: The Next Political Revolution," 838.
[8] See Ruger, "Global Health Governance," 654.

assistance.[9] Since global health institutions providing assistance characteristically focus on complying first and foremost with the donors' expectations, rather than with the affected communities' specific health needs and most urgent concerns, local communities are habitually neglected and excluded from the decision-making processes.

The centralization of decision-making power and authority in the hands of the WHO has typically been advocated as the proper remedy to both lack of coordination and lack of inclusion of local communities.[10] In centralizing more power and authority, the WHO would take on new roles. These would include, first, the responsibility of ensuring that both global health institutions providing assistance and the local communities receiving such assistance pursue a common goal. The idea here is to foster coordination by uniting the efforts of all stakeholders involved in a more complementary and efficient way. This effort would be done under the coordination guidance and leadership of the WHO. This means that the WHO would have a more active role in helping to define and clarify stakeholders' goals (individual and shared goals alike). Second, the WHO would take on the responsibility of ensuring that the health needs and most urgent concerns of local communities are prioritized over and above donors' expectations. The idea here is to empower local communities by fostering their inclusion and participation in healthcare resource allocation decision-making processes. This would be done under the centralizing power and authority of the WHO. The WHO would have a more active role in administrating all global public health funds, more specifically by concentrating the power to set priorities and allocate these resources accordingly.

The strength of the centralization approach lies in the fact that it reclaims, rather than shrinking, WHO's responsibilities as the global public health leader *par excellence*. However, it is not clear how centralization would more efficiently create better coordination among global health stakeholders, and how it would more effectively give voice and include marginalized communities into complex decision-making processes, where judgements need to be made in haste as an immediate response to the characteristic time-constraints of outbreaks. Instead of fostering efficient coordination and effective participation of local communities, centralization with the WHO will, I argue, have the opposite effect. First, centralization will expand WHO's bureaucracy, management costs, and therefore its inefficiency. Second, in further expanding the level of administration and governance structure, centralization in the hands of the WHO will only increase the distance from the local communities. Surely, if there

---

[9] See Gostin and Mok, "Innovative Solutions," 452.
[10] See Gostin et al., "The Normative Authority," 857; Mackey, "The Ebola Outbreak."

is more separation, there is inevitably a less effective dialogue. Coordination and inclusion, therefore, are not made easier but rather become more challenging as the administration and governance structure expand.

THE DECENTRALIZATION APPROACH

If the centralization approach is not a reasonable solution to the defects of global health governance, can decentralization be a more suitable alternative? To argue for the decentralization approach to global health governance, I introduce the principle of subsidiarity, on which the decentralization approach is predicated.

Origins and Development of Subsidiarity

The origins of the idea of subsidiarity can be traced back to classical Greece, where Aristotle (384–322 BCE) discusses the city as the community of communities, and the family, home, and village as the communal structures conducive to human flourishing. Subsequently, in the medieval period, Aquinas (1225–1274) further developed the idea of subsidiarity. Building on Aristotle's discussion on community, Aquinas elaborates on the idea of subsidiarity by explaining the dynamics among diverse groups composing a community. Then, in the sixteenth and seventeenth centuries, the Calvinist scholar Johannes Althusius (1557–1638) mentions the idea of subsidiarity within his theory of the federal state and his idea of "spheres of sovereignty." A myriad of political theorists followed Althusius and further echoed the idea. These theorists included John Locke (1632–1704), Montesquieu (1689–1775), Alexis de Tocqueville (1805–1859), John Stuart Mill (1806–1873), Pierre-Joseph Proudhon (1809–1865), and Abraham Lincoln (1809–1865).[11]

Later, in the nineteenth century, when the world wrestled with the two extremes of laissez-faire capitalism and Marxist socialism, Catholic social theorists, in seeking a more balanced approach, found a principled alternative in the idea of subsidiarity. It was in this historical context of the nineteenth and twentieth centuries that Catholic social thought substantially developed the idea of subsidiarity.[12] In 1891, Pope Leo XIII's *Rerum Novarum* highlighted the idea of subsidiarity. Although this papal encyclical did not explicitly use the term subsidiarity, the idea was clearly embedded in

---

[11] See Paolo G. Carozza, "Subsidiarity as a Structural Principle of International Human Rights Law," *American Journal of International Law* 97, no. 38 (2003): 38–79, at 41; Michelle Evans, "The Principle of Subsidiarity as a Social and Political Principle in Catholic Social Teaching," *Solidarity: The Journal of Catholic Social Thought and Secular Ethics* 3, no. 1 (2013): 44–60, at 44.

[12] On subsidiarity and Catholic social teaching, see also Lisa Sowle Cahill's chapter in this volume.

the text by stressing that "the State must not absorb the individual or the family; both should be allowed free and untrammeled action so far as is consistent with the common good and the interest of others. Rulers should, nevertheless, anxiously safeguard the community and all its members" (no. 35).

Although *Rerum Novarum* stressed the need for the State not to obliterate the individuals, the families, and the various other forms of smaller communities within its jurisdiction, the encyclical also sought to respond more directly to unfettered capitalism, advocating for public assistance and for an adequate protection of civil society associations, such as workers unions. As Pope Leo XIII wrote,

> The consciousness of his own weakness urges man to call in aid from without. It is this natural impulse which binds men together in civil society; and it is likewise this which leads them to join together in associations which are, it is true, lesser and not independent societies, but, nevertheless, real societies (*Rerum Novarum*, no. 50).

In 1931, the dynamic relationship between the larger society and the lesser societies was revisited in Pope Pius XI's *Quadragesimo Anno*. In acknowledging the societal changes and historical situation of the time, the papal encyclical draws attention to the unchanging character of the principle of subsidiarity, in that it orders the relations between small/lower-level and large/higher-level associations towards the common good of all:

> As history abundantly proves, it is true that on account of changed conditions many things which were done by small associations in former times cannot be done now save by large associations. Still, that most weighty principle, which cannot be set aside or changed, remains fixed and unshaken in social philosophy: Just as it is gravely wrong to take from individuals what they can accomplish by their own initiative and industry and give it to the community, so also it is an injustice and at the same time a grave evil and disturbance of right order to assign to a greater and higher association what lesser and subordinate organizations can do. For every social activity ought of its very nature to furnish help to the members of the body social, and never destroy and absorb them (no. 79).

If *Quadragesimo Anno* more directly responds to unfettered socialism, by emphasizing the negative component of subsidiarity (i.e., the negative duty of large/higher-level communities to refrain from interfering in the business of small/lower-level communities),[13] in 1961 Pope John XXIII would reiterate the positive component— duty of large/higher-level communities to assist small/lower-level

---

[13] See Pius XI, *Quadragesimo Anno*, no. 80.

communities—in *Mater et Magistra*. This papal encyclical explains subsidiarity as a remedy against the excesses of both public and private forms of ownership of property (no. 53). While an excessive public ownership of property may lead to a complete destruction of private ownership of property (no. 117), an excess of the latter ought also to be contained by the "public authority [which] must encourage and assist private enterprise, entrusting to it, wherever possible, the continuation of economic development" (no. 152).

Pope Saint John Paul II puts forth a similarly balanced approach in *Centesimus Annus*. Acknowledging the creation of "a new type of State, the so-called 'Welfare State'" (no. 48) as an attempt to answer and react to "many needs and demands, [and] forms of poverty and deprivation unworthy of the human person" (no. 48), *Centesimus Annus* recalls the principle of subsidiarity for the pursuit of an ordered structure, where large/higher-level communities and small/lower-level communities respect each other's spheres of competence and work together towards the common good. As Pope St. John Paul II puts it: "Here again *the principle of subsidiarity* must be respected: a community of a higher order should not interfere in the internal life of a community of a lower order, depriving the latter of its functions, but rather should support it in case of need and help to coordinate its activity with the activities of the rest of society, always with a view to the common good" (*Centesimus Annus*, no. 48).

Most recently, in 2005, Pope Benedict XVI's *Deus Caritas Est*, further stresses subsidiarity as a principle of justice. For a "just social order," Pope Benedict XVI clarifies that "We do not need a State which regulates and controls everything, but a State which, in accordance with the principle of subsidiarity, generously acknowledges and supports initiatives arising from the different social forces and combines spontaneity with closeness to those in need" (no. 26).

Catholic social scholars have extensively studied, debated, and therefore developed the idea of subsidiarity over the past centuries. This Catholic idea was adopted in 1992 as part of the EU law in the Treaty of Maastricht, to avoid excessive centralization of power and authority within the European system of governance and the EU community.[14] This is because the principle of subsidiarity allocates

---

[14] See Article 5(3) of the 1992 Maastricht Treaty on the European Union for the legal requirements of the principle of subsidiarity. The 2007 Treaty of Lisbon (a.k.a., Reform Treaty) amends the Maastricht Treaty. For the principle of subsidiarity, see article 5 of the Treaty of Lisbon: "1. The limits of Union competences are governed by the principle of conferral. The use of Union competences is governed by the principles of subsidiarity and proportionality.... 3. Under the principle of subsidiarity, in areas which do not fall within its exclusive competence, the Union shall act only if and insofar as the objectives of the proposed action cannot be sufficiently achieved by the Member States, either at central level or at regional and local level, but can

power and authority in a principled manner amongst the various levels of competence that exist in a certain community. It serves the purpose of ordering the governance structure of a community, and it does so by calling the problems in such community to be addressed from the bottom up, rather than the top-down perspective. Alternatively put, the principle of subsidiarity establishes that, only when the lowest sphere of power and authority proves ineffective in solving a certain problem, should the higher spheres of power step in, interfere, and become involved in those local affairs. To be sure, although the principle of subsidiarity does call for the intervention of higher spheres of power and the assistance (or *subsidium*) of the lower-level community when necessary, it equally calls for the respect and protection of the legitimate freedom and self-determination of the lower-level/small communities which are directly facing the problem.[15]

One reason for the allocation of power and authority in such bottom-up manner is efficiency: the principle of subsidiarity optimally distributes power and authority in a way that seeks to minimize waste, delay, and duplication. However, the principle of subsidiarity goes beyond the utilitarian rationale of efficiency. By concomitantly requiring (a) the larger (i.e., higher level) community's intervention and assistance reliant on (b) respect and protection of the legitimate freedom and self-determination of the smaller (i.e., lower level) community in need of assistance, the principle of subsidiarity entails coordination and inclusion.

First, the principle of subsidiarity necessitates coordination between the provider and the recipient of the assistance in such a way that their interests are harmonized. The provision of *subsidium* is only adequate if the needs and most urgent concerns of the smaller (i.e., lower level) community are met. Subsidiarity therefore requires that the best interests (i.e., the good) of the smaller (i.e., lower level) community are safeguarded. However, it also requires that the best interests (i.e., the good) of the larger (i.e., higher level) community, of which the smaller (i.e., lower level) community is part, is safeguarded as well. In other words, the principle of subsidiarity requires the good as well as the common good of the parts involved. Their interests

---

rather, by reason of the scale or effects of the proposed action, be better achieved at Union level. The institutions of the Union shall apply the principle of subsidiarity as laid down in the Protocol on the application of the principles of subsidiarity and proportionality. National Parliaments ensure compliance with the principle of subsidiarity in accordance with the procedure set out in that Protocol." European Parliament, "Treaty on the European Union/ Maastricht Treaty," www.europarl.europa.eu/about-parliament/en/in-the-past/the-parliament-and-the-treaties/maastricht-treaty.

[15] See Paolo G. Carozza, "The Problematic Applicability of Subsidiarity to International Law and Institutions," *American Journal of Jurisprudence* 61, no. 1 (2016): 51–67, at 51.

ought to be orchestrated, harmonized, and integrated into a coherent whole that upholds the flourishing of all (i.e., the common good).[16]

Second, the principle of subsidiarity is conditional upon the inclusion of the assisted community through their participation in the decisions pertaining to the assistance process. The provision of *subsidium* is only adequate if the legitimate freedom and self-determination of the smaller (i.e., lower level) community are respected, protected, and fulfilled. Subsidiarity therefore requires an appreciation for human agency, upon which all persons and communities need to act to realize and participate in their own good. Since human flourishing and the common good necessitate active participation of all persons and communities, rather than passive recipience of assistance, the principle of subsidiarity requires that the larger (i.e., higher level) community refrain from making decisions for the smaller (i.e., lower level) community. The smaller (i.e., lower level) community ought never to be treated as a mere passive recipient of material aid and ought to own her *subsidium*-related decisions, together with the larger (i.e., higher level) community.[17]

Subsidiarity and the Decentralization Approach to Global Health Governance

Both the centralization and the decentralization approaches share the goal of tackling global health governance's lack of coordination and inclusion. However, the means through which each approach proposes to achieve such common end differs substantially. On the one hand, the centralization approach advocates for a more powerful WHO, which concentrates more decision-making authority and therefore more responsibilities.[18] The decentralization approach, on the other hand, advocates for a more restrained WHO, which not only refrains from making decisions for the community in need of assistance—unless it is utterly incapable of doing so by itself—but also requires the assisted community to actively participate in all assistance-related decisions as a condition for the provision of aid. The decentralization approach may not be necessarily the fastest, as it gives the assisted community the time and space that it needs to make its own decisions and own its choices, which takes time and patience. Moreover, in PHEICs, since most decisions are often urgent, time is a particularly scarce and luxurious resource. The question of how much time the assisted community should be allowed to take is difficult and complex: for a prudent assessment, a number of factors (e.g., the spreadability of the PHEIC, its severity, resource-constraints, etc.) should be taken into consideration. However, the fact that this question

---

[16] See Carozza, "The Problematic Applicability," 53.
[17] See Carozza, "The Problematic Applicability," 54.
[18] See Mackey, "The Ebola Outbreak"; Gostin et al., "The Normative Authority," 857.

is difficult and complex does not mean that it is not ethically necessary. In dialogue with the community, it will be the task of global health policy makers to define a specific, prudent, and safe timeline.

Although the decentralization approach may not be the fastest and therefore the most efficient form of allocation of power and authority, the decentralization approach does minimize waste, delay, and duplication by requiring both coordination and inclusion. The interpretation of what coordination and inclusion mean within the centralization and the decentralization approaches varies. While the centralization approach understands "coordination" to mean that the WHO would have a more active role in helping define and clarify stakeholders' goals (individual and shared goals alike),[19] the decentralization approach invites the WHO to step back and leave the assisted communities to define their own goals as much as possible. The key idea behind the decentralization approach is that a good leader is a just leader. A just leader never steps in and interferes unnecessarily but rather respects the self-determination and protects the rhythm of development of those she assists with her leadership. In seeing the great leadership potential of the WHO, the decentralization approach would incentivize the WHO to lead by avoiding paternalistically imposing fixed goals or micromanaging the global health stakeholders under WHO coordination.

Likewise, while the centralization approach interprets "inclusion" to mean that the WHO would have a more active role in assuming the responsibility to administer all global public health funds—more specifically by concentrating the power to set priorities and allocate these resources accordingly[20]—the decentralization approach invites the WHO to assume less responsibilities. The key idea behind the decentralization approach is that a good leader is a wise and therefore humble leader who is not overcommitted and does not have overextended capacities. A good leader is prudently aware of her own limitations, weaknesses, and vulnerabilities and is ready to ask for assistance by delegating those tasks she will not be able to complete without aid. In recognizing the great leadership potential of the WHO, the decentralization approach would incentivize the WHO to lead by freeing itself from executing too many global health functions that could, instead, be executed by lower-level global health institutions, saving WHO's money, time, and energy, and allowing the WHO to exercise its core functions.

---

[19] See Mackey, "The Ebola Outbreak"; Gostin et al., "The Normative Authority," 857.
[20] See Tim K. Mackey and Bryan A. Liang, "A United Nations Global Health Panel for Global Health Governance," *Social Science and Medicine* 76, no. 1 (2013): 12–15, at 12–13.

CONCLUSION

It is commonly argued that the WHO should be reformed by way of centralizing more power and authority. Only then, so it is claimed, the WHO would be a good leader in global public health. I have challenged this mainstream view by arguing for a decentralized approach to global health governance, predicated on the EU law principle of subsidiarity. This principle justifies limitations to the roles and functions of higher-level structures of competence, such as the WHO. These limitations would free the WHO to focus on addressing the chief deficiencies of global health governance, namely lack of coordination and inclusion of local communities. The sort of coordination and inclusion proposed by the decentralization approach requires peculiar leadership skills—especially the ability to listen to those under one's leadership. This requires a leader with great disposition to be patient in hearing different voices, humble in learning by asking questions and including all under her leadership, and therefore wise in knowing when to assist or step back by respecting the persons and communities under her leadership.[21] M

Thana Cristina de Campos is assistant professor of Law, Ethics, and Public Policy at the Escuela de Gobierno of the Pontificia Universidad de Chile. She is also a Research Associate at the UNESCO Chair in Bioethics and Human Rights (Rome); the Von Hügel Institute (St. Edmund's College, University of Cambridge); and the Las Casas Institute (Blackfriars Hall, University of Oxford). She holds a DPhil in Law from the University of Oxford and an MPhil in International Law from the University of São Paulo. Dr. de Campos researches in global bioethics, international human rights, legal theory, political and moral philosophy, with a particular interest in natural law, virtue ethics, global health governance, and the human right to health. Besides numerous articles in prestigious journals, in 2017 she published the volume *The Global Health Crisis: Ethical Responsibilities*. Forthcoming is a co-edited book on *The Philosophical Foundations of Medical Law*.

---

[21] In this book, Alexandre Martins further stresses the importance of listening to the voices of those at the margins by fostering their engagement and by empowering them.

Part 5:
Building an Ethical Framework
For Education and Research
in Global Public Health

Inequities as an Ethical Imperative: Related to Identification, Engagement, Interventions in Minority Health

Nadia N. Abuelezam

WHEN QUANTIFYING AND CHARACTERIZING the health needs of populations in the United States, researchers often focus on differences between population subgroups to highlight how resources should be distributed and how health can be improved in the population. Medical and public health researchers are often asked to quantify differences to justify the distribution and allocation of resources and to report to agencies interested in understanding how health differs across the country. Subgroups that are of particular interest for the public health community are those that are underserved or under-resourced and those who may not belong to the "majority." These groups, herein referred to as minorities, include (but are not limited to) racial and ethnic minorities, sexual and gender minorities, rural populations, immigrant subgroups, and disabled individuals. In this chapter, I keep a broad definition of minority groups and provide examples across multiple population subgroups.

The focus on the difference in the public health and medical community is historically rooted in the origins of statistics and epidemiology. Statistics originated as a discipline interested in understanding the differences between White and non-White populations as a strategy to maintain power differentials and to justify unjust policies, including the enslavement of millions of Africans on United States territory.[1] The field of epidemiology had some of its origins in eugenics and aimed to highlight how White dominant populations had better health outcomes than non-White populations.[2] While the fields of statistics and epidemiology have aimed to move away from their eugenic origins and to improve the health of all

---

[1] See Nancy Krieger, *Epidemiology and the People's Health: Theory and Context* (New York: Oxford University Press, 2011).
[2] See Martin S. Pernick, "Eugenics and Public Health in American History," *American Journal of Public Health* 87, no. 11 (1997): 1767–1772.

populations, including those most vulnerable and marginalized, the history of these fields helps contextualize the obsession with difference and disparities.

Differences in health status can be categorized in one of two ways: disparities and inequities. Differences in health status or health outcomes between two groups are termed disparities. Differences that are routed in power imbalances, structural disadvantage, historical injustice, or other social justice issues are termed health inequities. While these terms are often used synonymously,[3] health inequities are those disparities that are rooted in a form of injustice and are often difficult to ameliorate or require structural changes to do so.

The field of public health and epidemiology has begun to address these challenges head on with attention being given to structural and social determinants of health in many subfields of public health.[4] Additionally, with the advent of social media and other forms of mass communication, communities in need have begun to document and advocate for their own health needs in society. Changes to the availability of data and the ability of social media to disseminate information widely have made disparities and inequities more widely known and discussed in the public health field.

Some of the starkest disparities currently observed in the American healthcare system involve maternal and infant health. Specifically, the maternal mortality rate in the United States has increased dramatically in recent years, and Black mothers and their infants have been experiencing the highest burden. Non-Hispanic Black infants have 2.3 times the infant mortality rate and are 3.8 times as likely to die from complications as non-Hispanic White infants.[5] There are also racial differences in the proportion of mothers who received prenatal care, with only two thirds of Black non-Hispanic mothers receiving prenatal care in the first trimester compared to 82.4 percent of non-Hispanic White mothers. Some works and writings have recently explored the connection of these statistics to historical injustices like the enslavement of Africans and the perceptions of pain and legitimacy of Black women among doctors and healthcare professionals.[6]

---

[3] See Olivia Carter-Pokras and Claudia Baquet, "What Is a 'Health Disparity'?" *Public Health Reports* 117, no. 5 (2002): 426–434.

[4] See Ana Penman-Aguilar et al., "Measurement of Health Disparities, Health Inequities, and Social Determinants of Health to Support the Advancement of Health Equity," *Journal of Public Health Management and Practice* 22, Suppl. 1 (2016): S33–S42.

[5] See Danielle M. Ely, and Anne K. Driscoll, "Infant Mortality in the United States, 2017: Data from the Period Linked Birth/Infant Death File," *National Vital Statistics Report* 68, no. 10 (2019): 1–19.

[6] See Dána-Ain Davis, "Obstetric Racism: The Racial Politics of Pregnancy, Labor, and Birthing," *Medical Anthropology* 38, no. 7 (2019): 560–573; Deirdre Cooper

Differences in access to healthcare and access to health insurance are also prevalent. Some of these disparities exist due to race, like the fact that, in 2018, Hispanic Americans are more than two times as likely to be uninsured as Whites.[7] Other issues with access and coverage are related to income and socioeconomic status, with individuals living below the poverty line being almost four times as likely to lack healthcare coverage.[8] Individuals living in rural areas are more likely to have low incomes and less likely to have health insurance covered by their employers.[9]

Recent evidence suggests that sexual and gender minority youth are at greater risk for mental health issues and violence, including suicide.[10] People living with disabilities are more likely to be obese, have cardiovascular disease, or not receive appropriate medical care due to cost than people without disabilities.[11]

DEMOGRAPHIC CHANGES

Public health is evolving constantly, and in the advent of large amounts of data on demographics, healthcare, and social engagement, more is now known about health and health needs than ever before. The changing demographic characteristics of American society also influence the degree of disparities. Primarily, the aging of the US population and the large amounts of immigration to the United States over the last century have changed the composition of the American population and the demands on the American healthcare system. The American population has diversified significantly with regard to race and ethnicity, creating many opportunities for engagement in understanding differences in health.

By 2050, it is estimated that the United States' population will increase by over fifty-three million and the majority of that growth will take place among people of color. While the proportion of Black

---

Owens, and Sharla M. Fett, "Black Maternal and Infant Health: Historical Legacies of Slavery," *American Journal of Public Health* 109, no. 10 (2019): 1342–1345.

[7] See Samantha Artiga et al., "Disparities in Health and Health Care: Five Key Questions and Answers," *Kaiser Family Foundation*, www.kff.org/disparities-policy/issue-brief/disparities-in-health-and-health-care-five-key-questions-and-answers/.

[8] See Artiga et al., "Disparities in Health and Health Care."

[9] See Vann Newkirk, and Anthony Damico, "The Affordable Care Act and Insurance Coverage in Rural Areas," *Kaiser Family Foundation*, www.kff.org/wp-content/uploads/2014/05/8597-the-affordable-care-act-and-insurance-coverage-in-rural-areas1.pdf.

[10] See Julia Raifman et al., "Sexual Orientation and Suicide Attempt Disparities among US Adolescents: 2009-2017," *Pediatrics* 145, no. 3 (2020): doi.org/10.1542/peds.2019-1658.

[11] See Gloria L. Krahn et al., "Persons with Disabilities as an Unrecognized Health Disparity Population," *American Journal of Public Health* 105, no. S2 (2015) S198–S206.

Americans is expected to stay the same, the proportion of Hispanic Americans and Asian Americans is expected to increase by forty-four percent and sixty percent respectively.[12] By 2050, it is estimated that the proportion of the population aged sixty-five or older will increase by ninety-six percent.[13] Recent evidence also shows that the proportion of adolescents reporting a sexual minority identity or same-sex sexual behavior has increased between 2009 and 2017 from 7.3 percent to 14.3 percent.[14] While these are only a few trends, they all point to the fact that changing demographics will lead to more diverse needs within the population over time.

Many of the differences being observed in American healthcare and health outcomes, as outlined above, are inherently related to historical injustices and structural issues. In order to ameliorate the inequities present in society, these structural determinants of health must be addressed head on. A more diverse American population, combined with our desire to understand differences in minority population subgroups, has its challenges. In particular, I aim to discuss the challenges related to the identification of health disparities and inequities, the engagement with minority subgroups, and the creation of solutions able to address health inequities in our changing American population.

IDENTIFICATION

One of the first steps in understanding and documenting health disparities and inequities is the ability to identify minority populations in the United States. Identification is a key aspect of public health as it helps detect risk factors and record health outcomes. There are many ways to identify minority groups. Systematic identification of groups through population-based records—including the census, birth certificates, and other national health records—would lead to the most accurate and systematic collection of data. While these data are ideal, they are not often available for all sub-groups of the population and may provide an incomplete description of population health. Large scale population-based surveys are also a good source of data but are limited by the types of questions asked and the ways in which these questions are asked. Standardized identity and sociodemographic questions on surveys are often dictated by government collection methods. The adoption of government survey questions makes identification a political action and requires stakeholders to adopt governmental language and preference. For example, a study on smoking among mothers found that the adoption of the Affordable

---

[12] See Artiga et al., "Disparities in Health and Health Care."
[13] See Centers for Disease Control and Prevention, "The State of Aging and Health in America 2013," www.cdc.gov/aging/pdf/state-aging-health-in-america-2013.pdf.
[14] See Raifman et al., "Sexual Orientation and Suicide."

Care Act's race and ethnicity categories masked differences in smoking among minority groups.[15] Additionally, when considering identifiers like race, it is important to remember that these groups are created by social scientists in order to better understand patterns, not because they best represent people's experiences.

Case Study: Arab Americans

An example of difficulty in identifying a minority subgroup is shown through work on Arab American health. Originally, when Arab immigrants began to immigrate to the United States, they were considered racially non-White despite being ethnically Arab.[16] Over time, Arab immigrants experienced discrimination and a denial of resources, including the right to vote, and advocated for a White racial designation. This White racial designation has persisted over time. Currently, the Office of Budget and Management considers individuals with historical origins from the Middle East and North Africa as White on standard racial surveys including the Census.[17] This racial designation has been shown not to fit universally with the lived experiences of this minority population, and there is evidence of increased rates of discrimination and stigma for Arab Americans over time.[18] Since the terrorist attacks on September 11, 2001, discrimination and stigma against Arab Americans has increased and has led to poorer mental health outcomes.[19] Additionally, categorizing Arab Americans as White prevents health researchers from properly assessing the health needs of this population,[20] despite some documented health disparities.[21] Without an ethnic identifier for this population, researchers are forced to use convenience samples and

---

[15] See Summer Sherburne Hawkins and Bruce B. Cohen, "Affordable Care Act Standards for Race and Ethnicity Mask Disparities in Maternal Smoking During Pregnancy," *Preventive Medicine* 65 (2014): 92–95.

[16] See Gregory Orfalea, *The Arab Americans: A History* (Northampton, MA: Olive Branch Press, 2006).

[17] See Executive Office of the President et al., "Standard for the Classification of Federal Data on Race and Ethnicity," www.whitehouse.gov/omb/fedreg_race-ethnicity.

[18] See Arab American Institute Foundation, "Underreported, under Threat: Hate Crime in the United States and the Targeting of Arab Americans 1991-2016," www.decodehate.com/reporthate.

[19] See Mona M. Amer and Joseph D. Hovey, "Anxiety and Depression in a Post-September 11 Sample of Arabs in the USA," *Social Psychiatry and Psychiatric Epidemiology* 47, no. 3 (2012): 409–418.

[20] See Nadia N. Abuelezam et al., "Arab American Health in a Racially Charged US," *American Journal of Preventive Medicine* 52, no. 6 (2017): 810–812.

[21] See Nadia N. Abuelezam et al., "The Health of Arab Americans in the United States: An Updated Comprehensive Literature Review," *Frontiers in Public Health* 6 (2018). doi.org/10.3389/fpubh.2018.00262.

other non-population based data sources to understand Arab American health.[22]

Recent work shows that Arab Americans do not have the same health needs as other immigrant and minority populations, establishing their unique healthcare needs and thus the need to be able to identify them in large population datasets.[23] While numerous scientists and activists have advocated for a unique Middle East and North African racial identifier, recently the federal administration denied the request to add a Middle East and North African identifier and indicated that this group will maintain their White racial designation.[24] This hinders the ability of public health researchers to appropriately understand the health needs of Arab Americans.[25] The slow political process of altering racial and ethnic categories on the Census is one example of how identification is crucial to a better understanding of minority health needs in the United States.

While the identification of Arab Americans is one example, many other minoritized groups are difficult to identify using standard surveillance methods, including sexual and gender minorities.[26] Identification of marginalized groups helps to create inclusive environments within healthcare and other social sectors. Being able to identify with a survey response category signals to the participants that they are included and valued in a study.[27] Researchers interested in advancing inclusion efforts in research studies must, therefore, pay attention to the ways in which minoritized groups are identified.

Identification dictates many aspects of knowledge production. It dictates what is known about groups and thus, cyclically, influences the funding available to better understand the health needs of particular groups and populations. If particular subgroups cannot be identified using standard identification strategies, especially those dictated by the federal government, federal funding agencies may not fund work to help these groups.[28] Identification then becomes the starting point for developing research programs in minority health.

---

[22] See Abuelezam et al., "Arab American Health."
[23] See Nadia N. Abuelezam et al., "Relevance of the 'Immigrant Health Paradox' for the Health of Arab Americans in California," *American Journal of Public Health* 109, no. 12 (2019): 1733–1738.
[24] See US Census Bureau, "Using Two Separate Questions for Race and Ethnicity in 2018 End-to-End Census," *2020 Census Program Memorandum Series*, www2.census.gov/programs-surveys/decennial/2020/program-management/memo-series/2020-memo-2018_02.pdf.
[25] See Abuelezam et al., "Arab American Health."
[26] See Joanne G. Patterson et al., "Measuring Sexual and Gender Minority Populations in Health Surveillance," *LGBT Health* 4, no. 2 (2017): 82–105.
[27] See Tahereh Moradi et al., "Translation of Questionnaire Increases the Response Rate in Immigrants: Filling the Language Gap or Feeling of Inclusion?" *Scandinavian Journal of Public Health* 38, no. 8 (2010): 889–892.
[28] See Abuelezam et al., "Arab American Health."

Accurate identification of minority groups is therefore crucial and necessary to identifying health needs in vulnerable and marginalized populations, to empowering these populations to engage with their health needs, and to creating appropriate interventions and solutions, as described in the next two sections.

ENGAGEMENT

While identification of subgroups is necessary, engagement with community members is required to truly understand the needs and desires of certain minoritized subgroups of the population. Specifically, research that prioritizes the experiences and voices of vulnerable populations, while also providing employment opportunities for those in the community, are ideal. Some argue that engagement with community members should be a precursor to identification as engagement could help define the ways in which community members can and want to be identified.[29] Engagement with communities is therefore an essential aspect of research in minority health.

The field of public health is plagued with examples of poor and unethical community engagement. Researchers and community health care workers are still trying to remedy the distrust in the public health and medical system that certain communities face. In particular, historically, African Americans have had a difficult and strained relationship with public health professionals. Many cite the Tuskegee Syphilis experiments, an unethical study performed on Black men with syphilis, as the catalyst to this relationship.[30] Evidence suggests, though, this strained relationship originated during experiments and poor medical treatment during mass enslavement.[31] Further, relationships between Native American populations and White populations have been strained over time, with health deteriorating on reservations over time.[32] Most recently, many well-known Black

---

[29] See Paulina O. Tindana et al., "Grand Challenges in Global Health: Community Engagement in Research in Developing Countries," *PLoS Medicine* 4, no. 9 (2007): doi.org/10.1371/journal.pmed.0040273.

[30] See Giselle Corbie-Smith, "The Continuing Legacy of the Tuskegee Syphilis Study: Considerations for Clinical Investigation," *American Journal of the Medical Sciences* 317, no. 1 (1999): 5–8; Dwayne T. Brandon et al., "The Legacy of Tuskegee and Trust in Medical Care: Is Tuskegee Responsible for Race Differences in Mistrust of Medical Care?" *Journal of the National Medical Association* 97, no. 7 (2005): 951–956.

[31] See Owens, and Fett, "Black Maternal and Infant Health: Historical Legacies of Slavery"; Vernellia R. Randall, "Slavery, Segregation and Racism: Trusting the Health Care System Ain't Always Easy—An African American Perspective on Bioethics," *Saint Louis University Public Law Review* 15, no. 2 (1995): 191–235.

[32] See B. Ashleigh Guadagnolo et al., "Medical Mistrust and Less Satisfaction with Health Care among Native Americans Presenting for Cancer Treatment," *Journal of Health Care for the Poor and Underserved* 20, no. 1 (2009): 210–226; Bonnie Duran et al., "Native Americans and the Trauma of History," in *Studying Native America:*

women have been speaking out about their experiences with maternal pain and morbidity during pregnancy, including tennis star Serena Williams.[33] The discussion of how and whether the pain and needs of Black women are prioritized in delivery settings has led to the discussion of distrust of medical professionals for this population.[34]

One of the reasons why community engagement is difficult is because the communities most disadvantaged have often been harmed or hurt historically by public health and medical professionals. This means that current public health professionals and clinicians must aim to take responsibility for past actions, make amends, and regain trust within these communities. It is essential that public health professionals work with communities to assess their needs and desires with regard to health. Working closely with communities to understand and right past harms may help improve relationships with communities in the future. Dismantling systems of oppression in society will go hand in hand with community engagement. Additionally, ensuring that healthcare professionals and public health researchers are trained in inclusive and anti-racist practices will ensure a more appropriate workforce that is attuned to the needs of individual communities.

Case Study: Program in Community Engagement

While public health has many examples of poorly handled community engagement, more recent and proactive strategies and methods—like community based participatory research—have helped engage communities in research in productive ways. The Program in Community Engagement has worked closely with Black and Latinx men who have sex with men (MSM) to ensure appropriate education around HIV prevention strategies. This program has been successful and longitudinal because of the community engagement aspects of the work. Example programs developed alongside community members include increased HIV testing, condom use, access to pre-exposure prophylaxis and antiretroviral therapy. The team and investigators behind these successful intervention programs have developed and outlined steps to ensure effective community building and partnership.

Problems and Prospects, ed. R. Thornton (Madison, WI: University of Wisconsin Press, 1998), 60–78.

[33] See Linda Villarosa, "Why America's Black Mothers and Babies Are in a Life-or-Death Crisis," *The New York Times Magazine*, April 11, 2018, www.nytimes.com/2018/04/11/magazine/black-mothers-babies-death-maternal-mortality.html.

[34] See Davis, "Obstetric Racism: The Racial Politics of Pregnancy, Labor, and Birthing."

These steps include community-based assessment, gathering feedback, and implementing interventions with community support.[35]

Engagement with communities requires a deep appreciation for a variety of cultures and languages, like the Program in Community Engagement. While it may be unreasonable to expect all public health professionals to be able to engage with all cultures effectively, diversifying the public health workforce to include members of minority communities will ensure the ability to engage with these communities effectively and respectfully. While the first step to this process is increasing representation of minority subgroups into public health education programs, additional work can be done to train local community members to participate in public health research and engage with the public health community. This can be achieved with methods developed to enhance community-based participatory research and patient-centered approaches to public health issues. The implementation of community health care workers in international settings is one example of community empowerment and engagement that has worked to reduce morbidity and mortality of diseases like HIV/AIDS and a variety of other infectious diseases.[36] While the presence of lay community healthcare workers in the United States is sparse, this could be an avenue of exploration to engage with and aid communities with limited resources or unique cultural perspectives on health.[37]

Programs aiming to increase representation in the public health workforce have implemented public health training earlier in students' research careers. Undergraduate programs in public health are flourishing across the country.[38] The interest and engagement of undergraduates in public health programs may help yield a more diverse set of public health workers in the future. Pipeline programs to increase representation in graduate programs are also common. These programs help students from minoritized backgrounds with stipends during graduate school as well as additional mentorship and support.[39] The effectiveness of these programs has not been examined

---

[35] See Scott D. Rhodes et al., "Engaged for Change: A Community-Engaged Process for Developing Interventions to Reduce Health Disparities," *AIDS Education and Prevention* 29, no. 6 (2017): 491–502.

[36] See Susan M. Swider, "Outcome Effectiveness of Community Health Workers: An Integrative Literature Review," *Public Health Nursing* 19, no. 1 (2002): 11–20.

[37] See Hector Balcazar et al., "Community Health Workers Can Be a Public Health Force for Change in the United States: Three Actions for a New Paradigm," *American Journal of Public Health* 101, no. 12 (2011): 2199–2203.

[38] See Beth Resnick et al., "An Examination of the Growing US Undergraduate Public Health Movement," *Public Health Reviews* 38, no. 1 (2017): doi.org/10.1186/s40985-016-0048-x.

[39] See Karen E. Bouye et al., "Increasing Diversity in the Health Professions: Reflections on Student Pipeline Programs," *Journal of Healthcare, Science and the Humanities* 6, no. 1 (2016): 67–79; Sonya G. Smith et al., "Pipeline Programs in the

longitudinally to assess overall impact on public health workforce diversification,[40] but having diverse students from different backgrounds will help bring culturally relevant and sensitive ideas to public health solutions and interventions.[41]

SOLUTIONS AND INTERVENTIONS

After identification of minority subgroups and engagement with community stakeholders, public health practitioners will be in a place to begin intervention work to help communities improve the health of their members. The solutions proposed, and work that is engaged in, must be structurally focused in order to make the largest impact on these communities and provide an ethical response. Structural interventions may have the largest and longest impact on these communities despite a general lack of clarity in the literature on how to implement these types of interventions. Structural solutions deal with existing structures like housing, employment, and education inequities in order to improve health.

Case Study: Communities of Opportunity

One of the most persistent and difficult power structures to address in American society is that of racism. Systematic racism alters the power structures and differentials for people of color and creates health disadvantages that can originate in utero.[42] In "Reducing Racial Inequities in Health," David Williams and Lisa Cooper have proposed a multi-level structural set of interventions that address systematic racism.[43]

The first step is to create local change through communities of opportunity.[44] Communities of opportunity are those that provide early access and engagement with children of low socioeconomic backgrounds in areas that have been historically disadvantaged. These

---

Health Professions, Part 1: Preserving Diversity and Reducing Health Disparities," *Journal of the National Medical Association* 101, no. 9 (2009): 836–851.

[40] See Jonathan Fuchs et al., "Growing the Pipeline of Diverse HIV Investigators: The Impact of Mentored Research Experiences to Engage Underrepresented Minority Students," *AIDS and Behavior* 20, no. 2 (2016): 249–257.

[41] See Darcell P. Scharff and Matthew W. Kreuter, "Training and Workforce Diversity as Keys to Eliminating Health Disparities," *Health Promotion Practice* 1, no. 3 (2000): 288–291.

[42] See Cheryl L. Giscombé and Marci Lobel, "Explaining Disproportionately High Rates of Adverse Birth Outcomes among African Americans: The Impact of Stress, Racism, and Related Factors in Pregnancy," *Psychological Bulletin* 131, no. 5 (2005): 662–683.

[43] See David R. Williams and Lisa A. Cooper, "Reducing Racial Inequities in Health: Using What We Already Know to Take Action," *International Journal of Environmental Research and Public Health* 16, no. 4 (2019): doi.org/10.3390/ijerph16040606.

[44] See Williams and Cooper, "Reducing Racial Inequities in Health."

communities ensure access to early high-quality childhood development resources and provide work and income support for parents. Housing and neighborhood conditions are also monitored and regulated in these communities to ensure adequate and appropriate environments in which children can grow. This work aims to create change and opportunity at the local community level. The second step is to engage with health care professionals and the healthcare system to ensure that prevention is emphasized as a guiding principle in the medical care being provided to these communities.[45] This step addresses concerns related to healthcare quality that is responsive to the needs of individual communities. The third step is to engage with policymakers to disseminate knowledge and information about the health inequities that are occurring in these communities to help inform the public and garner support.[46]

The structural approach to communities of opportunity ensures that public health officials are addressing systematic issues at multiple levels and provides a comprehensive response for communities that have been marginalized. It directly addresses the social determinants of health that are leading to poorer health outcomes among those in minority groups.[47]

Structural responses to health disparities and inequities require an understanding of the context and an authentic engagement with communities.[48] Additionally, interventions should be specific to timing and location and should not be replicable universally.[49] Challenges to responding structurally include a lack of rigorous study designs, ineffective and inefficient methods, and limited funds. Changing structural elements of society and communities may have unintended consequences that affect more than just health in these areas.[50] Because much of the research in this realm has been biomedical in nature, understanding how best to disseminate results and information to stakeholders is a challenge.

Structural solutions are often different from the typical biomedical solutions public health practitioners are used to implementing. Because of this, schools of public health and other institutions must be

---

[45] See Williams and Cooper, "Reducing Racial Inequities in Health."
[46] See Williams and Cooper, "Reducing Racial Inequities in Health."
[47] See Rachel L.J. Thornton et al., "Evaluating Strategies for Reducing Health Disparities by Addressing the Social Determinants of Health," *Health Affairs* 35, no. 8 (2016): 1416–1423.
[48] See Kim M. Blankenship et al., "Structural Interventions: Concepts, Challenges and Opportunities for Research," *Journal of Urban Health* 83, no. 1 (2006): 59–72.
[49] See David A. Dzewaltowski et al., "Behavior Change Intervention Research in Community Settings: How Generalizable Are the Results?" *Health Promotion International* 19, no. 2 (2004): 235–245.
[50] See Kim M. Blankenship et al., "Structural Interventions in Public Health," *AIDS* 14, Suppl. 1 (2000): S11–S21.

intentional about training practitioners in these methods.[51] This will likely mean that faculty trained in methods related to social and structural determinants of health work would be prioritized in new faculty hires, increasing the diversity of thought in many institutions. It is through a change in engagement with social and structural determinants of health that change can be made in schools of public health and medical schools and the communities they serve.

CONCLUSION

While the health and medical fields have long been fascinated by differences, it is up to the next generation of public health leaders and professionals to engage with these differences and develop structural solutions that could create change. Historical injustices and systems of oppression that influence health must be targeted directly by healthcare professionals and public health practitioners. We must not shrink away from our responsibility to improve the public's health through active and forward-thinking identification, respectful and empowering community engagement, and structurally-based solutions and interventions to health problems. The future of minority health requires a new vocabulary and way of thinking about public health and its ramifications in society. M

Nadia N. Abuelezam, ScD, is assistant professor at the Connell School of Nursing at Boston College and an epidemiologist. She has expertise in biostatistics, mitigating health inequities for minority health, and data analytic approaches in public health. Her current research focuses on immigrant health and, particularly, on women's health and mental health outcomes. She relies on quantitative methods and novel data streams to better understand the inequalities in health care distribution and access in resource-poor settings and among vulnerable populations.

---

[51] See Helena Hansen, and Jonathan M. Metzl, "New Medicine for the US Health Care System: Training Physicians for Structural Interventions," *Academic Medicine: Journal of the Association of American Medical Colleges* 92, no. 3 (2017): 279–281.

Journal of Moral Theology

Addressing Health Disparities among Families: Policy Approaches to Improve Infant Health

Summer Sherburne Hawkins

THE BURDEN OF POOR HEALTH FALLS UPON the most vulnerable families. Parents with low educational attainment and low household income, often referred to as measures of socioeconomic status (SES), are more likely to engage in poor health behaviors that are the leading causes of disease burden, including higher intake of sugar-sweetened beverages (SSBs), levels of heavy drinking and tobacco use, lower intake of fruits and vegetables, and lower levels of physical activity compared to their more advantaged counterparts.[1] These maladaptive health behaviors often translate into worse health outcomes, such that adults from disadvantaged households are more likely to be overweight or obese or have heart disease, and are at higher risk of mortality.[2] Children

---

[1] See Colin D. Rehm et al., "Dietary Intake among US Adults, 1999-2012," *JAMA* 315, no. 23 (2016): 2542–2553; Adam M. Leventhal et al., "Association of Cumulative Socioeconomic and Health-Related Disadvantage with Disparities in Smoking Prevalence in the United States, 2008 to 2017," *JAMA Internal Medicine* 179, no. 6 (2019): 777–785; Camillia K. Lui et al., "Educational Differences in Alcohol Consumption and Heavy Drinking: An Age-Period-Cohort Perspective," *Drug and Alcohol Dependence* 186 (2018): 36–43; Grainne O'Donoghue et al., "Socio-Economic Determinants of Physical Activity across the Life Course: A 'Determinants of Diet and Physical Activity' (DEDIPAC) Umbrella Literature Review," *PLoS One* 13, no. 1 (2018): doi.org/10/1371/journal.pone.0190737; William M. Schultz et al., "Socioeconomic Status and Cardiovascular Outcomes: Challenges and Interventions," *Circulation* 137, no. 20 (2018): 2166–2178.

[2] See Ricardo A. Pollitt et al., "Evaluating the Evidence for Models of Life Course Socioeconomic Factors and Cardiovascular Outcomes: A Systematic Review," *BMC Public Health* 5 (2005): doi.org/10.1186/1471-2458-5-7; Cynthia L. Ogden et al., "Prevalence of Obesity among Adults, by Household Income and Education: United States, 2011–2014," *Morbidity and Mortality Weekly Report* 66, no. 50 (2017): 1369–1373; Barry Bosworth, "Increasing Disparities in Mortality by Socioeconomic Status," *Annual Review of Public Health* 39 (2018): 237–251; Silvia Stringhini et al., "Socioeconomic Status and the 25 x 25 Risk Factors as Determinants of Premature Mortality: A Multicohort Study and Meta-Analysis of 1.7 Million Men and Women," *Lancet* 389, no. 10075 (2017): 1229–1237.

from disadvantaged households are also at higher risk of poor health. Infants born to mothers with low SES are more likely to be born with low birth weight or preterm,[3] be overweight across childhood,[4] be exposed to secondhand smoke,[5] and become smokers themselves during adolescence.[6]

The intergenerational transmission of disadvantage, as described by Cheng and colleagues, suggests that children born into poverty are more likely to become adults with low SES and suffer the associated health burdens.[7] Breaking this cycle requires improving families' economic circumstances to help both generations escape poverty. This approach focuses on high-risk families with evidence demonstrating longer-term improvements in children's cognitive, emotional, and behavioral outcomes.[8] Taken together, this suggests that improving the health behaviors and outcomes of parents has important downstream consequences for the health and well-being of their children and subsequent generations.

It is essential to recognize that race/ethnicity in the United States is also an important determinant of health. Although most health behaviors and outcomes are racially/ethnically patterned,[9] educational and income-based disparities often cut across race/ethnicity. As will be described in Section 4, women with low levels of education have the highest prevalence of smoking for both white and Black women despite low-educated white women being twice as likely to smoke

---

[3] See Philip Blumenshine et al., "Socioeconomic Disparities in Adverse Birth Outcomes: A Systematic Review," *American Journal of Preventive Medicine* 39, no. 3 (2010): 263–272.

[4] See T. D. Brisbois et al., "Early Markers of Adult Obesity: A Review," *Obesity Reviews* 13, no. 4 (2012): 347–367; Cynthia L. Ogden et al., "Prevalence of Obesity among Youths by Household Income and Education Level of Head of Household: United States 2011–2014," *Morbidity and Mortality Weekly Report* 67, no. 6 (2018): 186–189.

[5] See David M. Homa et al., "Vital Signs: Disparities in Nonsmokers' Exposure to Secondhand Smoke—United States, 1999–2012," *Morbidity and Mortality Weekly Report* 64, no. 4 (2015): 103–108; Katherine King et al., "Family Composition and Children's Exposure to Adult Smokers in Their Homes," *Pediatrics* 123, no. 4 (2009): doi.org/10.1542/peds.2008-2317.

[6] See Darren Mays et al., "Parental Smoking Exposure and Adolescent Smoking Trajectories," *Pediatrics* 133, no. 6 (2014): 983–991; Judith S. Brook et al., "The Intergenerational Transmission of Smoking in Adulthood: A 25-Year Study of Maternal and Offspring Maladaptive Attributes," *Addictive Behaviors* 38, no. 7 (2013): 2361–2368.

[7] See Tina L. Cheng et al., "Breaking the Intergenerational Cycle of Disadvantage: The Three Generation Approach," *Pediatrics* 137, no. 6 (2016): doi.org/10/1542/peds.2015.2467.

[8] See Cheng et al., "Breaking the Intergenerational Cycle of Disadvantage."

[9] See Paula A. Braveman et al., "Socioeconomic Disparities in Health in the United States: What the Patterns Tell Us," *American Journal of Public Health* 100, Suppl. 1 (2010): S186–S196.

compared to low-educated Black women. This chapter focuses more on disparities as measured by SES, income, and education than racial/ethnic differences. Other scholars have provided excellent reviews of the literature on how social forces, such as racism, can induce biological changes that result in adverse health outcomes.[10]

Health behaviors and outcomes are not innately different across families, but these differences are influenced by the social determinants of health. The World Health Organization (WHO) has defined the social determinants of health as the "conditions in which people are born, grow, work, live, and age, and the wider set of forces and systems shaping the conditions of daily life."[11] Examples of the social determinants of health are: limited access to education, employment that is not protected by smoke-free legislation, neighborhoods that have a high volume of liquor outlets, living in a food desert, limited access to opportunities for engaging in healthy behaviors, and lack of protection from unhealthy exposures. A social justice approach identifies upstream factors that can be modified to increase opportunities and support healthy choices for vulnerable populations rather than blame individuals. Section 4 presents a case study of how cigarette taxes are an upstream approach that has been shown to reduce disparities in prenatal smoking.

The contexts in which families are embedded impacts the health of its members across the life course and intergenerationally. Frameworks have been developed to conceptualize the multiple determinants of population health, depicting the influential roles of individual factors, social and community networks, living and working conditions, social, economic, health, and environmental conditions, and policies on population health.[12] Interventions at any of these 'layers' can influence these contexts and alter the health of parents and their children. With a social justice lens, the goal is not only to improve the health of all families but to pay particular attention to the most vulnerable. Because some policies have been shown to widen health disparities,[13] it is critical to evaluate their impact on both

---

[10] See Earnestine Willis et al., "Conquering Racial Disparities in Perinatal Outcomes," *Clinics in Perinatololgy* 41, no. 4 (2014): 847–875; David R. Williams et al., "Understanding How Discrimination Can Affect Health," *Health Services Research* 54, Suppl. 2 (2019): 1374–1388.
[11] World Health Organization, "Social Determinants of Health," www.who.int/social_determinants/en/.
[12] See Institute of Medicine, *The Future of the Public's Health in the Twenty-First Century* (Washington, DC: National Academies Press, 2002); Göran Dahlgren and Margaret Whitehead, *Policies and Strategies to Promote Social Equity in Health* (Stockholm: Institute for the Futures Studies, 1991).
[13] See Katherine L. Frohlich and Louise Potvin, "Transcending the Known in Public Health Practice: The Inequality Paradox: The Population Approach and Vulnerable Populations," *American Journal of Public Health* 98, no. 2 (2008): 216–221.

the entire population of interest as well as the gap between the most and least disadvantaged. Health equity will be achieved when the most vulnerable in our society achieve the same level of health as the most advantaged.

This chapter focuses on the impact of policies enacted at the outermost layers of these determinants of population health frameworks on health disparities among families. The majority of such policies and programs are enacted to directly influence the health of vulnerable children, such as Head Start,[14] which has been shown to positively influence health, social, and educational outcomes among the most vulnerable children. However, less is known about policies that may indirectly influence children's health via impacting the health or economic circumstances of mothers or parents more broadly. For example, legislation may be enacted to allow unconventional oil and natural gas development, like fracking, to support economic growth and reduce the reliance on foreign energy supplies; however, research has since shown that prenatal exposure to fracking is associated with low birth weight and preterm birth.[15] Therefore, it is critical to evaluate the downstream effects of policies on parental health behaviors and child health outcomes because they may have unintentional consequences that would otherwise remain unknown.

TYPES OF POLICIES INFLUENCING POPULATION HEALTH

There are multiple types of policies that can directly or indirectly influence the engagement in health behaviors and, subsequently, population health. Four such policies are legislation, regulation, recommendations, and fiscal policies. First, *legislation* is laws enacted that describe policy statements or objectives with information on who and what is to be governed and the procedures and means of enforcement.[16] In this context, legislation is related to a health issue either through means of health promotion by protecting rights of engagement or inhibiting engagement through punitive consequences. Second, *regulations* are rules made by and maintained by an

---

[14] See Michael Puma et al., *Third Grade Follow-up to the Head Start Impact Study Final Report, OPRE Report # 2012-45* (Washington, DC: Office of Planning, Research and Evaluation, Administration for Children and Families, US Department of Health and Human Services, 2012).

[15] See Janet Currie et al., "Hydraulic Fracturing and Infant Health: New Evidence from Pennsylvania," *Science Advances* 3, no. 12 (2017): doi.org/10/1126/sciadv.16032; Kristina Walker Whitworth et al., "Drilling and Production Activity Related to Unconventional Gas Development and Severity of Preterm Birth," *Environmental Health Perspectives* 126, no. 3 (2018): doi.org/10.1289/EHP2622; Joan A. Casey et al., "Unconventional Natural Gas Development and Birth Outcomes in Pennsylvania, USA," *Epidemiology* 27, no. 2 (2016): 163–172.

[16] See World Health Organization, "Legislation and Regulation," www.who.int/heli/tools/legis_regul/en/.

authoritative body or regulatory agency and enforced by law. Regulations set standards or administrative requirements for the implementation of a law related to a health issue.[17] Third, *recommendations* are evidence-based guidelines made by the government or authoritative bodies to engage or not engage in specific health behaviors. Such recommendations are suggestive to support positive health behaviors or reduce maladaptive health behaviors but are without penalty. Fourth, *fiscal* policies are the means by which governments adjust their spending in the form of taxes and subsidies. Taxes increase price to discourage the engagement in unhealthy behaviors as well as to generate revenues for the government. Subsidies reduce price and support the purchase of health promoting and disease or injury preventing products or services. Tobacco control provides an illustrative case study of each of these policy approaches.

Tobacco Control Policies

Smoke-free legislation was first created to protect employees from secondhand smoke exposure in the workplace.[18] The legislation prohibited the use of any combustible tobacco product in a workplace, restaurant, or bar and has been since extended to other settings such as gambling establishments, college campuses, cars, and even outdoor venues such as public parks.[19] While research has demonstrated that secondhand smoke exposure has decreased as a direct result of smoke-free legislation and this decrease has population-level benefits, including reductions in cardiovascular events, its effect on smoking prevalence and tobacco consumption is inconsistent.[20] A more recent example of tobacco legislation occurred in December 2019, when the President signed legislation that raised the federal minimal age of the sale of any tobacco product, including e-cigarettes, to twenty-one years.[21]

In 2009, the Family Smoking Prevention and Tobacco Control Act gave the US Food and Drug Administration (FDA) authority to regulate the manufacturing, marketing, and sale of tobacco products.[22]

---

[17] See World Health Organization, "Legislation and Regulation."
[18] See Andrew Hyland et al., "Smoke-Free Air Policies: Past, Present and Future," *Tobacco Control* 21, no. 2 (2012): 154–161.
[19] See American Nonsmokers' Rights Foundation, "U.S. Tobacco Control Laws Database," no-smoke.org/materials-services/lists-maps/.
[20] See Kate Frazer et al., "Legislative Smoking Bans for Reducing Harms from Secondhand Smoke Exposure, Smoking Prevalence and Tobacco Consumption," *Cochrane Database of Systematic Reviews* 2 (2016): doi.org/10.1002/14651858.CD005992.pub3.
[21] See US Food and Drug Administration, "Tobacco 21," www.fda.gov/tobacco-products/retail-sales-tobacco-products/tobacco-21.
[22] See US Food and Drug Administration, "Family Smoking Prevention and Tobacco Control Act: An Overview," www.fda.gov/tobacco-products/rules-regulations-and-guidance/family-smoking-prevention-and-tobacco-control-act-overview.

This authority regulates the design and safety of conventional and new tobacco products and requires companies to test and report on all tobacco product constituents, ingredients and additives, by brand.[23] As the FDA notes, "The fact that FDA regulates these tobacco products does not mean they are safe to use."[24] Despite the regulation of tobacco products, the role of public health and health professionals is to ensure that the public receives accurate information about the health consequences of conventional and emerging tobacco products.[25]

Recommendations made by the government, governing bodies, or public health organizations aim to provide evidence-based guidance on health topics. The US Preventive Services Task Force (USPSTF) is an independent panel of experts which makes evidence-based recommendations about clinical preventive services. The USPSTF has published recommendations on tobacco smoking cessation in adults.[26] The USPSTF recommends that clinicians ask all adults, including pregnant women, about tobacco use and advise them to stop using tobacco. Non-pregnant adults should be offered behavioral interventions and FDA-approved pharmacotherapy for cessation, while pregnant women should be provided with behavioral interventions only. The USPSTF has also concluded that there is insufficient evidence to recommend electronic nicotine delivery systems for tobacco cessation in adults, including pregnant women.[27] As specifically related to families, the American College of Obstetricians and Gynecologists and the American Academy of Pediatrics both recommend not to smoke during pregnancy or after delivery.[28]

---

[23] See Tobacco Control Legal Consortium, *Federal Regulation of Tobacco: A Summary* (Saint Paul, MN: Tobacco Control Legal Consortium, 2009).

[24] US Food and Drug Administration, "Products, Ingredients & Components," www.fda.gov/tobacco-products/products-guidance-regulations/products-ingredients-components.

[25] See National Center for Chronic Disease Prevention and Health Promotion Office on Smoking and Health, *The Health Consequences of Smoking—50 Years of Progress: A Report of the Surgeon General* (Atlanta, GA: Centers for Disease Control and Prevention, 2014).

[26] See US Preventive Services Task Force, "Tobacco Smoking Cessation in Adults, Including Pregnant Women: Behavioral and Pharmacotherapy Interventions," www.uspreventiveservicestaskforce.org/Page/Document/UpdateSummaryFinal/tobacco-use-in-adults-and-pregnant-women-counseling-and-interventions1.

[27] See US Preventive Services Task Force, "Tobacco Smoking Cessation."

[28] See Committee on Underserved Women and Committee on Obstetric Practice, "Committee Opinion No. 721: Smoking Cessation During Pregnancy," *Obstetrics & Gynecology* 130, no. 4 (2017): doi.org/10.1097/AOG.0000000000002353; Harold J. Farber et al., "Clinical Practice Policy to Protect Children from Tobacco, Nicotine, and Tobacco Smoke," *Pediatrics* 136, no. 5 (2015): 1008–1017.

Taxes to raise the cost of manufacturing, distributing, retailing, or consuming unhealthy products[29] are one of the most effective policies to prohibit the engagement in unhealthy behaviors, including tobacco use.[30] Public health taxes, often referred to as 'sin taxes', target behaviors with an element of choice; cigarettes, alcohol, and SSBs are not basic needs and individuals can choose to consume them or not.[31] Evidence suggests that consumption of these substances is sensitive to price. Based on the law of demand in economics, when the price of a product increases, demand for that product decreases. Research suggests that increasing the price of cigarettes by ten percent reduces consumption by three to six percent.[32] However, this is an average change and research has demonstrated that low-income households are more price sensitive. Increases in the price of these products have a larger effect among low-income households and, consequently, they consume fewer of these products.

Research has shown that lower SES adults are more likely to engage in unhealthier behaviors governed by public health taxes than their higher SES counterparts.[33] Some scholars have argued that since public health taxes are regressive—i.e., low-income earners pay a larger percentage of their income on tobacco taxes than high-income earners—they are unethical because the tax burden is not shared equally across the economic spectrum.[34] Since the price elasticity of these products is low—meaning product substitution is minimal—price increases will reduce the available income in lower income households to purchase essential items. In contrast, other scholars have argued that since there is a higher burden of tobacco-related illness and death among lower SES populations, then these same groups will have larger gains in health from tax increases that

---

[29] See Alexandra Wright et al., "Policy Lessons from Health Taxes: A Systematic Review of Empirical Studies," *BMC Public Health* 17, no. 1 (2017): 583.
[30] See World Health Organization, *WHO Report on the Global Tobacco Epidemic, 2015: Raising Taxes on Tobacco* (Geneva: World Health Organization, 2015).
[31] See World Health Organization, "Public Health Taxes," www.who.int/health_financing/topics/sin-taxes/en/.
[32] See Frank J. Chaloupka and Kenneth E. Warner, "The Economics of Smoking," in *Handbook of Health Economics 1B*, eds. Anthony J. Culyer and Joseph P. Newhouse, Handbooks in Economics (Amsterdam: Elsevier, 2000), 1539–1628; International Agency for Research on Cancer, "Tax, Price and Aggregate Demand for Tobacco Products," in *Effectiveness of Tax and Price Policies for Tobacco Control* (Lyon: International Agency for Research on Cancer, 2011).
[33] See Leventhal et al., "Association of Cumulative Socioeconomic and Health-Related Disadvantage;" Rehm et al., "Dietary Intake among US Adults;" Lui et al., "Educational Differences in Alcohol Consumption;" Schultz et al., "Socioeconomic Status and Cardiovascular Outcomes."
[34] See Dahlia K. Remler, "Poor Smokers, Poor Quitters, and Cigarette Tax Regressivity," *American Journal of Public Health* 94, no. 2 (2004): 225–229.

encourage reducing consumption or cessation.[35] The WHO endorses public health taxes, and specifically tobacco taxes,[36] as an effective strategy to reduce consumption of products that adversely affect health, reduce the burden on the health system as a result of decreased health care needs, and increase revenue for public health programs and the health sector.[37]

IMPACT OF TAXES ON INFANT HEALTH

As illustrated by tobacco control, there are multiple types of policy tools that can be utilized to promote the health of populations and families as well as reduce disparities. However, many of these policies are targeted at adults with less consideration about the potential downstream effects on their offspring. Low birth weight and preterm birth are associated with poor health across the life course.[38] Thus, the goal should be increasing birth weight into the normal range (2.5–4.5 kg [5.5–9.9 lbs]), reducing babies born low birth weight (<2.5 kg [5.5 lbs]), and encouraging delivery at term (i.e., forty weeks).

Consumption of both alcohol and SSBs conforms to the law of demand and is sensitive to price. Research has shown that increasing the price of alcohol by ten percent reduces consumption by approximately 5 percent.[39] Taxation on SSBs is a more recent policy, but estimates suggest that a ten percent increase in price leads to a decreased consumption of eight to ten percent.[40] Both of these policies are primarily aimed at adults, as alcohol cannot be legally purchased until age twenty-one and the purchase of SSBs is often by adults in a household. While taxes on SSBs are a promising area of research to help address rising levels of obesity,[41] their effect on birth outcomes

---

[35] See M. Scollo and E.M. Greenhalgh, "Are Tobacco Taxes Regressive?," in *Tobacco in Australia: Facts and Issues*, eds. M. M. Scollo and M. H. Winstanley (Melbourne: Cancer Council Victoria, 2019), www.tobaccoinaustralia.org.au/chapter-13-taxation/13-11-are-tobacco-taxes-regressive; Remler, "Poor Smokers, Poor Quitters, and Cigarette Tax Regressivity."

[36] See World Health Organization, *WHO Report on the Global Tobacco Epidemic, 2015*.

[37] See World Health Organization, "Public Health Taxes."

[38] See David J.P. Barker and Kent L. Thornburg, "The Obstetric Origins of Health for a Lifetime," *Clinical Obstetrics and Gynecology* 56, no. 3 (2013): 511–519.

[39] See Alexander C. Wagenaar et al., "Effects of Beverage Alcohol Price and Tax Levels on Drinking: A Meta-Analysis of 1003 Estimates from 112 Studies," *Addiction* 104, no. 2 (2009): 179–190.

[40] See Tatiana Andreyeva et al., "The Impact of Food Prices on Consumption: A Systematic Review of Research on the Price Elasticity of Demand for Food," *American Journal of Public Health* 100, no. 2 (2010): 216–222.

[41] See Wright et al., "Policy Lessons from Health Taxes: A Systematic Review of Empirical Studies," Yilin Yoshida and Eduardo J. Simoes, "Sugar-Sweetened Beverage, Obesity, and Type 2 Diabetes in Children and Adolescents: Policies, Taxation, and Programs," *Current Diabetes Reports* 18, no. 6 (2018): doi.org/10.1007/s11892-018-1004-6.

is unknown. In addition, only one study has examined the impact of alcohol taxes on infant health. There is no safe level of alcohol consumption during pregnancy, and the American College of Obstetricians and Gynecologists advises mothers not to drink alcohol while pregnant.[42] Alcohol travels in the mother's bloodstream through the placenta to the developing fetus and has been associated with low birth weight.[43] Using data from natality files in 1985–2002, Zhang found that a one-cent increase in beer taxes raised the average birth weight by 1 gram, in wine taxes 0.2–0.3 grams, and in liquor taxes 0.7 grams.[44] Similarly, a one-cent increase in beer taxes decreased the incidence of low birth weight by 0.1–2 percentage points. Despite known differences in alcohol consumption across women's SES, this study did not examine the impact of alcohol taxes on disparities in birth outcomes. Unfortunately, follow-up studies are not possible as alcohol use during pregnancy is no longer collected on the birth certificate.

CASE STUDY: CIGARETTE TAXES

An area that has received greater attention is the impact of tobacco control policies, particularly cigarette taxes, on disparities in smoking during pregnancy. In 2017, in the US, 17.7 percent of women reported smoking prior to pregnancy, with 8.1 percent continuing into the third trimester of pregnancy.[45] However, these are average levels, and my colleagues and I have shown there are striking disparities in prenatal smoking. Using data from the Pregnancy Risk Assessment Monitoring System (PRAMS)—a state-representative survey of women postpartum—there is over a twelve percentage point gap in the prevalence of prenatal smoking between women with the lowest and highest educational attainment (13.5 percent versus 1.1 percent) and household income (16.7 percent versus 1.1 percent).[46] White women are also more likely to report smoking during pregnancy (10.0 percent)

---

[42] See American College of Obstetricians and Gynecologists, "FAQ: Fetal Alcohol Spectrum Disorders (FASDs)," www.acog.org/About-ACOG/ACOG-Departments/Tobacco--Alcohol--and-Substance-Abuse/Fetal-Alcohol-Spectrum-Disorders-Prevention-Program/FAQs?IsMobileSet=false.

[43] See Loubaba Mamluk et al., "Low Alcohol Consumption and Pregnancy and Childhood Outcomes: Time to Change Guidelines Indicating Apparently 'Safe' Levels of Alcohol During Pregnancy? A Systematic Review and Meta-Analyses," *BMJ Open* 7, no. 7 (2017): doi.org/10.1136/bmjopen-2016-015410.

[44] See Ning Zhang, "Alcohol Taxes and Birth Outcomes," *International Journal of Environmental Research and Public Health* 7, no. 5 (2010): 1901–1912.

[45] See Centers for Disease Control and Prevention, "Prevalence of Selected Maternal and Child Health Indicators for All PRAMS Sites, Pregnancy Risk Assessment Monitoring System (PRAMS), 2016-2017," www.cdc.gov/prams/prams-data/mch-indicators/states/pdf/2018/All-PRAMS-Sites-2016-2017_508.pdf.

[46] See Summer Sherburne Hawkins et al., "Use of ENDS and Cigarettes During Pregnancy," *American Journal of Preventive Medicine* 58, no. 1 (2020): 122–128.

than Black (9.0 percent) or Hispanic (2.5 percent) women. Using women's race/ethnicity and educational attainment together identifies women at the highest risk of prenatal smoking. Among white women with 0–11 years of education, 31.5 percent reported smoking during pregnancy compared to 4.5 percent of women with thirteen or more years of education, with corresponding levels among Black women as 16.7 percent versus 4.9 percent, respectively. In contrast, between 2.3-2.5 percent of Hispanic women report smoking during pregnancy across all levels of educational attainment.

Using data from the birth certificate, including over twenty-six million births across forty-seven states and D.C., we have shown that the average prevalence of smoking during the first trimester of pregnancy has decreased from 11.5 percent to 7.9 percent from 2005 to 2015.[47] Prenatal smoking is known to be under-reported on the birth certificate compared to the PRAMS survey administered a few months after delivery,[48] suggesting these figures are likely under-estimates of the true prevalence. Racial/ethnic and education disparities in smoking reported during the first trimester are evident in these data as well. As seen in Figure 1, women's educational attainment is protective against smoking during pregnancy for both white and Black women, but the gradient is much steeper for the former than the latter. There is over a thirty percentage point gap in prenatal smoking between white women with the lowest and highest levels of education while this gap is over twelve percentage points among Black women.

---

[47] See Summer Sherburne Hawkins and Christopher F. Baum, "The Downstream Effects of State Tobacco Control Policies on Maternal Smoking During Pregnancy and Birth Outcomes," *Drug and Alcohol Dependence* 205 (2019): doi.org/10.1016/j.drugalcdep.2019.107634.

[48] See Van T. Tong et al., "Trends in Smoking before, During, and after Pregnancy—Pregnancy Risk Assessment Monitoring System, United States, 40 Sites, 2000-2010," *Morbidity and Mortality Weekly Report Surveillance Summary* 62, no. 6 (2013): 119.

Figure 1. Prevalence (percent) of maternal smoking during pregnancy by race/ethnicity and educational attainment as reported on the US certificate of live birth.[49]

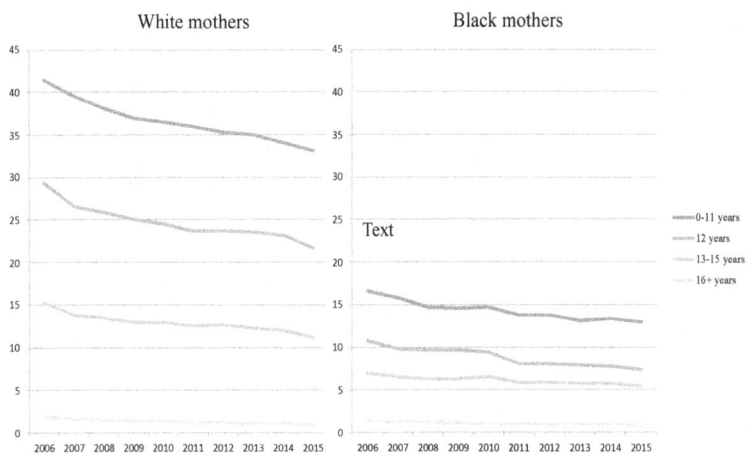

Birth outcomes also follow a strong educational gradient. Education is protective for all women, such that babies born to high-educated white and Black women weigh two hundred grams more than babies born to their low-educated counterparts.[50] Furthermore, babies born to low-educated mothers are nearly twice as likely to be born low birth weight or preterm. However, it is also important to note that Black women, on average, have poorer birth outcomes than white women across the educational spectrum. Taken together, the most disadvantaged women have the highest prevalence of smoking during pregnancy and their babies have the poorest birth outcomes.[51]

Cigarette smoking is a modifiable determinant of low birth weight.[52] The detrimental health consequences of prenatal use of cigarettes are well-established, with infants of smokers weighing 150

---

[49] Figure 1 is based on data presented in Hawkins and Baum, "The Downstream Effects."
[50] See Hawkins and Baum, "The Downstream Effects."
[51] See Hawkins and Baum, "The Downstream Effects."
[52] See US Department of Health and Human Services, *How Tobacco Smoke Causes Disease—The Biology and Behavioral Basis for Smoking-Attributable Disease: A Report of the Surgeon General: Executive Summary* (Rockville, MD: US Department of Health and Human Services, 2010).

to 200 grams less than infants of nonsmokers. These infants also are more likely to be born low birth weight and preterm. Nicotine crosses the placenta, and in utero exposure has been associated with adverse neurocognitive outcomes, insufficient lung development, and compromised lung function. Secondhand smoke exposure has also been shown to affect birth outcomes. Infants born to non-smoking pregnant women exposed to secondhand smoke are, on average, twenty-eight grams lighter and more likely to be born preterm.[53] Therefore, reducing cigarette consumption or quitting smoking altogether by parents and family members have important implications for infant health.

The US has experienced widespread cigarette tax increases and the enactment of smoke-free legislation over last two decades. As of January 2020, the average cigarette tax was $1.81 per pack, but this ranged from $0.17 in Missouri to $4.50 in Washington, D.C.[54] In addition, twenty-seven states have enacted 100 percent smoke-free legislation in workplaces, restaurants, and bars.[55] Since cigarette taxes and smoke-free legislation are enacted at the state-level, this has created a natural experiment of policy changes within and across states over time. The question remains: can cigarette tax increases and smoke-free legislation reduce disparities in prenatal cigarette use and improve birth outcomes?

Evaluations of Tobacco Control Policies on Birth Outcomes

My colleagues and I have conducted a series of studies to test whether strengthening tobacco control policies across US states have reduced prenatal smoking and improved birth outcomes among the most vulnerable infants. It is important to note that the surveillance of maternal smoking during pregnancy on the birth certificate has changed over the past 30 years. The 1989 revision of the birth certificate asked mothers to report whether they smoked during pregnancy (yes/no) and, separately, the average number of cigarettes per day. In contrast, the 2003 revision of the birth certificate asked mothers to report the average number of cigarettes smoked per day

---

[53] See National Center for Chronic Disease Prevention and Health Promotion Office on Smoking and Health, *The Health Consequences of Smoking—50 Years of Progress*; US Department of Health and Human Services, *How Tobacco Smoke Causes Disease—The Biology and Behavioral Basis for Smoking-Attributable Disease*; Lucinda J. England et al., "Nicotine and the Developing Human: A Neglected Element in the Electronic Cigarette Debate," *American Journal of Preventive Medicine* 49, no. 2 (2015): doi.org/10.1016/j.amepre.2015.01.015.
[54] See Campaign for Tobacco-Free Kids, "Map of State Cigarette Tax Rates," www.tobaccofreekids.org/research/factsheets/pdf/0222.pdf.
[55] See American Nonsmokers' Rights Foundation, "U.S. 100% Smokefree Laws in Non-Hospitality Workplaces and Restaurants and Bars," www.no-smoke.org/pdf/WRBLawsMap.pdf.

during each trimester of pregnancy. The Centers for Disease Control and Prevention (CDC) has stated that data from these two questions are not comparable.[56]

We first used natality files based on the 1989 revision of the birth certificate from 2000–2010 on nearly eighteen million births from twenty-eight states and Washington, D.C. We found that for every $1.00 increase in cigarette taxes, low-educated white and Black mothers decreased smoking by nearly two percentage points and smoked between fourteen and twenty-two fewer cigarettes per month.[57] Linking changes in smoking to birth outcomes, we then found that the birth weight of babies born to white mothers increased by over five grams and babies born to Black mothers by four grams.[58] Among these mothers, cigarette tax increases also reduced the risk of having babies born low birth weight and preterm.

More recently, we used the natality files based on the 2003 revision of the birth certificate from 2005–2015 on over 26 million infants from 47 states and Washington, D.C., to examine the impact of more recent policy changes on disparities in prenatal smoking and birth outcomes.[59] We found that low-educated white and Black women continue to be the most responsive to cigarette tax increases. Every $1.00 increase in taxes was associated with a 3.45 percentage point decrease in prenatal smoking among white mothers and a 1.20 percentage point decrease among Black mothers. These declines in prenatal smoking translated into increases in birth weight by 4.19 grams for babies born to white mothers and 0.89 grams for babies born to Black mothers. We also found some evidence that tax increases reduced the likelihood of babies born with low birth weight and preterm.[60] In contrast, in both sets of studies we found no effect of the enactment of smoke-free legislation on prenatal smoking or birth outcomes.[61]

In sum, state cigarette tax increases had the largest benefits for the most vulnerable mothers and infants. Cigarette taxes reduced smoking during pregnancy among low-educated white and Black women and,

---

[56] See Centers for Disease Control and Prevention, "State Cigarette Minimum Price Laws: United States, 2009," *Morbidity and Mortality Weekly Report* 59, no. 13 (2010): 389–392.
[57] See Summer Sherburne Hawkins and Christopher F. Baum, "Impact of State Cigarette Taxes on Disparities in Maternal Smoking During Pregnancy," *American Journal of Public Health* 104, no. 8 (2014): doi.org/10.2105/AJPH.2014.301955.
[58] See Summer Sherburne Hawkins et al., "Associations of Tobacco Control Policies with Birth Outcomes," *JAMA Pediatrics* 168, no. 11 (2014): doi.org/10.1001/jamapediatrics.2014.2365.
[59] See Hawkins and Baum, "The Downstream Effects."
[60] See Hawkins and Baum, "The Downstream Effects."
[61] See Hawkins et al., "Associations of Tobacco Control Policies with Birth Outcomes"; Hawkins and Baum, "Impact of State Cigarette Taxes"; Hawkins and Baum, "The Downstream Effects."

as a result, their babies were more likely to have higher birth weights and less likely to be born with low birth weight or preterm. While improvements in birth weight by a few grams may appear trivial, they have substantial implications at the population level. As noted in our paper, if Missouri increased its tax to that of D.C., moving from the lowest in the country to the highest, then the "birth weight of infants born to Missouri mothers with less than a high school degree would increase, on average, by 18.1 grams for white infants and 3.9 grams for Black infants."[62] Improvements in birth outcomes are welcome byproducts of state-level policies aimed at adult health behaviors.

LESSONS LEARNED AND IMPLICATIONS

The most vulnerable families in the US and globally bear the highest burden of poor health. The intergenerational transmission of poverty means that disadvantage and subsequent health implications will continue unless there are interventions to break that cycle.[63] State policies are one type of intervention aimed at the outermost level in determinants of population health frameworks.[64] As illustrated by tobacco control, there are many policy responses to shape the context in which families are embedded—legislation, regulation, recommendations, and fiscal policies—and all have the potential to influence the social determinants of health.

My colleagues and I have contributed to the body of evidence demonstrating that fiscal policies, particularly taxes, have downstream effects on infant health by influencing women's health behaviors during pregnancy. Because mothers with low SES have the highest prevalence of poor health behaviors and their babies have the worst birth outcomes,[65] policies have the potential to reduce that gap. Our research in tobacco control has demonstrated that cigarette taxes are an effective policy tool to reduce smoking during pregnancy among women with the highest levels of smoking and improve the birth outcomes of their infants.[66] Further areas to consider are how to increase state taxes that are lower than the national average, whether

---

[62] Hawkins and Baum, "The Downstream Effects."
[63] See Cheng et al., "Breaking the Intergenerational Cycle of Disadvantage: The Three Generation Approach."
[64] See Institute of Medicine, *The Future of the Public's Health*; Dahlgren and Whitehead, *Policies and Strategies*.
[65] See Hawkins and Baum, "The Downstream Effects"; Cheryl Robbins et al., "Disparities in Preconception Health Indicators–Behavioral Risk Factor Surveillance System, 2013-2015, and Pregnancy Risk Assessment Monitoring System, 2013-2014," *Morbidity and Mortality Weekly Report Surveillance Summary* 67, no. 1 (2018): 116.
[66] See Hawkins and Baum, "The Downstream Effects"; Hawkins et al., "Associations of Tobacco Control Policies with Birth Outcomes"; Hawkins and Baum, "Impact of State Cigarette Taxes."

there is a ceiling effect for cigarette taxes, and at what point do tax increases take resources away for essential household items, particularly among at-risk families. Currently, these questions remain unanswered.

While the prevalence of smoking during pregnancy is lower in other regions of the world than the US and Europe,[67] tobacco use among adolescent girls is higher in many countries than among adult women.[68] This suggests the prevalence of smoking during pregnancy may increase in the future as this younger cohort reaches reproductive age. Furthermore, the global coverage of WHO's EMPOWER package of tobacco control policies is unequally distributed.[69] Despite WHO's recommendation to implement cigarette taxes because they are the most effective policy to reduce tobacco use,[70] only 14 percent of the world population is covered by taxation.[71] Tobacco taxes are one of the most well-researched fiscal policies, and more studies are needed to examine the implications of other public health taxes and subsidies on population health.

In conclusion, it is essential that the unintended consequences of policies are considered during the political process and evaluations are conducted to understand their ramifications. While the focus of this chapter is on families and infants, unintended consequences could also befall the elderly, immigrants, or other groups that have been historically disenfranchised. If the downstream effects of policies are not examined then their consequences simply may not be identified. In order to improve the health and well-being of the most vulnerable children and families, political decisions need to be made by taking a longer-term, evidence-based approach to population health. ∎

Summer Sherburne Hawkins, PhD, MS, joined the Boston College School of Social Work (BCSSW) faculty in 2012. She is a social epidemiologist interested in addressing policy-relevant research questions in maternal and child health. Her research examines the impact of policies on health disparities in parents and children, particularly using methodology that integrates epidemiology and economics. Dr. Hawkins published on the topics of parental and adolescent tobacco use, infant feeding practices, and childhood obesity as well as the impact of state and local policies on disparities in these health behaviors and outcomes. A more recent area of her research focuses on the role of the Affordable Care Act on the uptake of

---

[67] See Shannon Lange et al., "National, Regional, and Global Prevalence of Smoking During Pregnancy in the General Population: A Systematic Review and Meta-Analysis," *Lancet Global Health* 6, no. 7 (2018): doi.org/10.1016/S2214-109X(18)30223-7.

[68] See Tobacco Atlas, "Youth," tobaccoatlas.org/topic/youth/.

[69] See World Health Organization, *WHO Report on the Global Tobacco Epidemic* (Geneva: World Health Organization, 2019).

[70] See World Health Organization, "Public Health Taxes."

[71] See World Health Organization, *WHO Report on the Global Tobacco Epidemic*.

preventive health services. In January 2017, she received a three-year grant from the American Cancer Society (ACA) titled "Impact of the ACA on the prevention and early detection of women's cancers." Dr. Hawkins is currently on the editorial board of the journal *Maternal and Child Nutrition* and an active member of the University Research Council at Boston College. Prior to joining BCSSW, Dr. Hawkins was a Cohort 7 Robert Wood Johnson Health & Society Scholar at the Harvard School of Public Health.

Journal of Moral Theology

Humanitarian Aid, Infectious Diseases, and Global Public Health

Nils Hennig

THIS CHAPTER EXAMINES ETHICAL CHALLENGES in humanitarian aid, infectious diseases, and global public health. It describes key tasks in emergency situations, ethical issues in global health work, and trends in neglected infectious diseases. It provides the relevant ethical principles needed to develop an ethical framework. The principles can be applied to particular situations, guiding desperately needed action in global health practice. The chapter ends with some words of caution and final thoughts on humanitarian aid and global health.

THE TOP TEN PRIORITIES OF INTERVENTION

The ethical challenges in humanitarian aid are plentiful, especially in humanitarian emergencies when we have to respond to multiple priorities with often very limited resources. Refugee crises are not a new problem but date back to the earliest days of humanity. Humanitarian organizations have developed experience-based priorities when assisting refugees. The necessary interventions in an emergency phase usually cover ten top priorities, which should be carried out simultaneously. They have to be implemented rapidly and attention must be given not just to quantity but quality. These ten top priorities include: (1) initial assessment, (2) water and sanitation, (3) food and nutrition, (4) shelter and site planning, (5) health care in the emergency phase, (6) control of communicable diseases and epidemics, (7) measles immunization, (8) public health surveillance, (9) human resources and training, and (10) coordination.[1]

The initial assessment allows an informed decision on whether or not to intervene and identifies intervention priorities. It should be completed within three days, using simple, straightforward methods, and should result in quick decisions.[2] The mortality rate is often the

---

[1] See Medécins sans Frontières, *Refugee Health: An Approach to Emergency Situations* (London: Macmillan Education, 1997), 38.
[2] See Medécins sans Frontières, *Refugee Health*, 53.

best indicator for assessing the severity of a situation and should be established as a base-line for evaluating the effectiveness of an intervention. Security conditions must be clearly described since they can have limiting effects on the presence of intervention teams and affect the implementations of programs.

Water and sanitation play an essential role in the spread of communicable diseases and epidemics. In an emergency situation, priority must be given to meet drinking and cooking needs and basic hygiene requirements.[3] Human excreta (mostly fecal matter) are always contaminated and must be disposed of in well-defined areas. Proper personal and environmental hygiene prevent vectors from developing and spreading disease. Dealing with the dead, which is often essential for infection control, can lead to social tensions with the community. During the recent Ebola epidemic in West Africa efforts to eliminate traditional mourning and burial rituals led to distrust and much worse.[4]

Food and nutrition intervention addresses basic food needs and decreases mortality and morbidity resulting from malnutrition. An assessment of the food and nutritional situation should always be part of the initial health assessment. The prevalence of acute malnutrition in children less than five years of age can generally be used as an indicator of this condition in the entire population since this group is more sensitive to changes in the nutritional situation.[5] The food provided has to be adequate in quantity and nutrient content. The objectives are to treat severely and moderately malnourished persons and to prevent malnutrition in vulnerable groups. The main factors required for successful and regular distribution are political willingness, adequate planning of the food supply, registration of the population, good organization of the distribution, and regular monitoring.[6] A major obstacle during conflicts is inequity in access and food diversion by powerful and often armed groups, especially if these groups are in control of distribution.

Early shelter and site planning minimize overcrowding and make it possible to organize efficient relief services. Priorities to take into consideration include security, access to water, environmental health risk, and the local population. Cultural and social patterns should be taken into account.

---

[3] See Médecins sans Frontières, *Refugee Health*, 79.
[4] See Junerlyn Agua-Agum et al., "Exposure Patterns Driving Ebola Transmission in West Africa: A Retrospective Observational Study," *PLoS Medicine* 13, no. 11 (2016): doi.org/10.1371/journal.pmed.1002170. See also Jacquineau Azetsop's chapter in this book.
[5] See Médecins sans Frontières, *Refugee Health*, 111.
[6] See Médecins sans Frontières, *Refugee Health*, 89.

Healthcare in the emergency phase has to be flexible; it must adapt to the evolving situation and changing needs. Whenever possible, existing facilities should be used and supported. However, in most camp situations, new services have to be set up. Health services should focus on basic curative care.

Communicable diseases are the primary cause of mortality among displaced populations and have to be controlled. Preventive measures are most effective and outbreaks require a specific response. The four diseases responsible for most mortality include diarrheal disease, respiratory infections, measles, and malaria.[7] By affecting children, measles continues to be one of the most severe health problems in the world. Controlling measles includes an immunization campaign. The recommended strategy is the organization of a first and rapid mass campaign coupled with vitamin A supplementation, to be followed by a routine immunization program integrated within existing health facilities.

A surveillance system is essential to provide early warnings of epidemics. Data collection should be simple and limited to public health problems that can and will be acted upon; crude mortality rate is most useful to measure the gravity of the situation.[8]

Human resources and training should follow specific procedures and be supervised by experienced staff. Labor laws of the host country have to be considered. Training will be necessary, but it should be preceded by an assessment of training needs. Proper coordination is essential and, when neglected, an intervention will often become disastrous. The involvement of local stakeholders is key, and information exchange should be formalized.

ETHICAL ISSUES IN GLOBAL HEALTH WORK

Humanitarian aid workers constantly face ethical challenges during the emergency phase. I became aware of these challenges while working in Angola with the Doctors Without Borders/Médecins sans Frontières (MSF)—an international humanitarian group dedicated to providing medical care to people in distress. The conflict in Angola started before independence in 1975. The main players were UNITA (National Union for the Total Independence of Angola), a powerful, totalitarian guerilla movement supported by the United States, on one side, and the MPLA (Popular Movement for the Liberation of Angola), a repressive Marxist-Leninist party-state, allied and supported by Cuba and the Soviet Union initially, and Angola's oil

---

[7] See Medécins sans Frontières, *Refugee Health*, 152.
[8] See Medécins sans Frontières, *Refugee Health*, 204.

wealth at the time, on the other side.[9] Between 1998 and 2002, the last years of the conflict were especially gruesome and violent.[10] At the end of the civil war in Angola, we suddenly had access to the battle zones and rebel areas without assistance or protection, the so-called "gray zones." Tens of thousands of Angolans, unable to find food, perished in these gray zones during the last years of the war and the first weeks of peace. During our initial assessment, we encountered thousands of severely malnourished civilians, too weak to move, waiting to die. The numbers were overwhelming. For the first days and weeks, we had to decide whom to take with us for treatment and whom to leave behind for certain death. We identified hundreds of patients who needed immediate treatment in our feeding centers, but only had very limited transportation capacity. What guides a decision on whom to leave behind to perish? Can an ethical framework be applied in this situation?

What are some of the main ethical issues and conflicts encountered in global public health work? The goals of public health include promoting health and preventing disease and achieving the ethical and human rights principle of health equity. Health equity means that everyone has a fair and just opportunity to be as healthy as possible. This requires removing obstacles to health such as poverty, discrimination, and their consequences, including powerlessness and lack of access to good jobs with fair pay, quality education and housing, safe environments, and health care.[11] Therefore, reducing and eliminating health disparities is fundamental to reaching health equity. Considerations of justice are central to global health. The global health care status quo reflects a collective failure of the international community to meet the most basic needs of the world's population.[12] Lack of global justice and solidarity, combined with corruption in the public and private sector, are major hurdles. As the example from Angola demonstrates, the injustice of limited resource allocation has to be addressed.

Often the problem of limited access to health care in resource-poor countries is exacerbated by a "brain drain": the loss of trained professional personnel to wealthier countries that offers greater opportunity and pay. The issue of "brain drain" raises additional ethical issues regarding the acceptability of such recruitment and of

---

[9] See Christine Messiant, "Angola: Woe to the Vanquished," in *In the Shadow of 'Just Wars': Violence, Politics and Humanitarian Action*, ed. Fabrice Weissman (London: C. Hurst, 2004), 109–136, at 111.
[10] See Messiant, "Angola: Woe to the Vanquished," 118.
[11] See Paula Braveman et al., *What Is Health Equity? And What Difference Does a Definition Make?* (Princeton, NJ: Robert Wood Johnson Foundation, 2017).
[12] See World Health Organization, *Global Health Ethics: Key Issues* (Geneva: World Health Organization, 2015), 19.

actions to hinder migration.[13] The conflict is between the freedom to relocate and associate freely on one side and the need to improve the health of vulnerable people on the other. How does global public health navigate the relationship between the liberty of the individual and broader societal concerns? Infectious disease outbreaks—like the COVID-19 pandemic—threaten the health and welfare of others, and it may be legitimate to restrict people's liberty in order to protect the community. But how far should authorities be allowed to go in the name of the "greater good" of disease control? Or would the refusal to implement strict travel restrictions, cancel social gatherings and implement quarantine signal a lack of solidarity with the most vulnerable?

Global health action needs good data. How can the needs for accurate surveillance be balanced against the principle of individual autonomy?[14] Should individuals be tested for a disease when providers are unable to offer them appropriate medications due to poor resource allocation but the collected data might allow access to medications for the community in the future? Testing might help the community but can bring devastation to the individual. I have seen persons who tested positive for HIV being ostracized by their community, but it was data from these tests that eventually convinced the international community to provide HIV treatment for resource-poor countries. Would it have been more or less ethical not to provide the individuals with their test results (as was also often done)?

Another set of ethical issues arises when assigning foreign workers for deployment during emergencies. As a strict rule, foreign aid workers should be deployed only if they are capable of providing necessary services not sufficiently available in the local setting.[15] Assignment of outside health workers has to take into consideration their relevant skills and knowledge as well as their linguistic and cultural competencies.[16] It is inappropriate to deploy unqualified or unnecessary workers mainly to satisfy a personal or professional desire to be "helpful." Following the tsunami in Southeast Asia, there was an unprecedented outpouring of international aid and humanitarian workers.[17] But being on site and having worked in Banda Aceh in the days and months after the tsunami, it became obvious to me that a large number of foreign aid workers were ill-prepared and

---

[13] See World Health Organization, *Global Health Ethics*, 19. See also Daniel J. Daly's chapter in this volume.
[14] See World Health Organization, *Global Health Ethics*, 15.
[15] See World Health Organization, *Guidance for Managing Ethical Issues in Infectious Disease Outbreaks* (Geneva: World Health Organization, 2016), 47.
[16] See World Health Organization, *Guidance for Managing Ethical Issues*, 48.
[17] See Richard M. Zoraster, "Barriers to Disaster Coordination: Health Sector Coordination in Banda Aceh Following the South Asia Tsunami," *Prehospital and Disaster Medicine* 21, Suppl. 1 (2006): S13–S18.

their presence had more to do with disaster tourism than humanitarian aid. Any foreign aid workers deployed during crises, especially where resources are scarce, should carefully consider (before departure!) whether they are prepared to deal with ethical issues that may lead to moral and psychological distress.[18]

Finally, we have to acknowledge that people in different countries and societies may hold different values or place different weights on common values. Cultural relativism is the idea that a person's beliefs, values, and practices should be understood based on that person's own culture, rather than be judged against the criteria of another. Applying global health ethics may lead to accusations of ethical imperialism. In a refugee camp in Sierra Leone, one very early morning, I surprised some of the staff performing female genital mutilation on a young girl in the clinic. My intervention to stop the procedure led to the mutilation being done with unsterile instruments somewhere else and to mistrust from the local staff which saw the mutilation as an integral part of their culture. Still, while some might argue that condemning female genital mutilation and other practices as human rights violations constitute an ill-advised form of ethical imperialism, others argue that we must stand up for the women and children who are at risk of being harmed.[19]

Global health ethics should protect individuals and the public from harm and promote the highest attainable standard of care. Issues of standards of care in resource-poor settings are a real practical concern. All of these ethical issues are made worse during natural disasters, armed conflict, and infectious disease outbreaks.

GLOBAL TRENDS IN INFECTIOUS DISEASES AND NEGLECTED INFECTIOUS DISEASES

What are global trends in infectious diseases? After having devastated humankind for most of history and keeping human life expectancy well below forty years of age, infectious diseases receded in Western countries in the twentieth century due to urban sanitation, improved housing, personal hygiene, and vaccination.[20] Antibiotics further suppressed morbidity and mortality. But since the last quarter of the twentieth century, we see new and resurgent infectious diseases.[21] Examples of new infectious diseases include HIV, Lyme disease, Lassa Fever, Nipah Virus, H1N1 influence, SARS, MERS-CoV, and COVID-19. Examples of re-emerging/resurging infectious

---

[18] See World Health Organization, *Guidance for Managing Ethical Issues*, 48.
[19] See World Health Organization, *Global Health Ethics*, 20.
[20] See Joshua Lederberg, "Infectious History," *Science* 288, no. 5464 (2000): 287–293.
[21] See David M. Morens et al., "The Challenge of Emerging and Re-Emerging Infectious Diseases," *Nature* 430, no. 6996 (2004): 242–249.

diseases include cholera, the plague, dengue, yellow fever, Chikungunya fever, West Nile fever, and multiple drug resistant/extensively drug resistant (MDR/XDR) tuberculosis. According to the WHO, infectious diseases are spreading more rapidly than ever before, and new infectious diseases are being discovered at a higher rate than at any time in history.[22]

Especially in low-income countries, neglected infectious diseases continue to cause significant morbidity and mortality. Yet, of the eight hundred fifty new therapeutic products approved between 2000 and 2011, only four percent (and only one percent of all approved New Chemical Entities) were indicated for neglected diseases, even though these diseases account for eleven percent of the global disease burden.[23] Although some of these illnesses are finally getting the priority that is necessary to control or even eradicate them, others are still barely recognized except by the individuals who suffer from them. Selected examples include sleeping sickness, leishmaniasis, Chagas, pediatric HIV, filarial diseases, hepatitis C, and mycetoma.

Sleeping sickness or Human African Trypanosomiasis is endemic in thirty-six African countries and about sixty-five million people are at risk of being infected.[24] Sleeping sickness is transmitted by the tsetse fly and is fatal without treatment. Over 1 billion people are at risk of leishmaniasis worldwide.[25] The parasite that leads to infection is called Leishmania and transmitted by sandflies. Existing treatments are difficult to administer, toxic, and costly. Drug resistance is also an increasing problem. Chagas disease is endemic in twenty-one Latin American countries, where it kills more people than malaria. In total, seventy million people are at risk worldwide. Less than one percent of patients have access to treatment.[26] 1.7 million children below fifteen years of age are living with HIV globally, mainly in sub-Saharan Africa.[27] Three hundred of them die every day. Filarial diseases—such as lymphatic filariasis (elephantiasis), onchocerciasis (river blindness), and loiasis (loa loa)—cause chronic illness and life-long disabilities leading to great suffering and social stigmatization. Over twenty-one million people are infected with river blindness alone, and

---

[22] See World Health Organization, *World Health Report 2007—A Safer Future: Global Public Health Security in the 21st Century* (Geneva: World Health Organization, 2007).
[23] See Belen Pedrique et al., "The Drug and Vaccine Landscape for Neglected Diseases (2000–11): A Systematic Assessment," *Lancet Global Health* 1, no. 6 (2013): doi.org/101016/S2214-109X(13)70078-0.
[24] See Drugs for Neglected Diseases initiative, "Diseases and Projects," https://www.dndi.org/diseases-projects/.
[25] See Drugs for Neglected Diseases initiative, "Diseases and Projects."
[26] See Drugs for Neglected Diseases initiative, "Diseases and Projects."
[27] See Drugs for Neglected Diseases initiative, "Diseases and Projects."

two hundred five million people are at risk.[28] Seventy-one million people worldwide are chronically infected with hepatitis C. Seventy-five percent of them live in low- and middle-income countries. Effective medicines are now available, but their high cost means that only thirteen percent of hepatitis C patients globally have access to treatment.[29] Mycetoma is a stigmatizing disease often resulting in devastating deformities, amputation, and morbidity. The exact route of infection is unknown. Treatment success for eumycetoma is less than thirty-five percent.[30] An ethical framework would not only include the goal that these diseases will cease to be neglected but that society will also cease to neglect the people suffering from them. New partnerships, leading to innovative research and to developing therapies for these diseases, are needed.

KEY ETHICAL ISSUES IN GLOBAL HEALTH RESEARCH

That brings us to some key ethical issues in global health research. Global health research aims to improve lives by testing existing and new treatments, preventive measures, and systems and procedures. It has produced great public health benefits, but it has also been the cause of concerns.[31] Key ethical questions that have to be addressed include: Does the research have social value for the communities that take part or from which the participants are drawn? Who benefits from the research?[32] When studies are carried out in resource-low settings, the individuals who take part and are put at risk may not be able to benefit from any knowledge gained by the study due to their poor economic status. Often the research agenda is driven by the potential profit and commercial success of new drugs and devices. Due to market forces, the pharmaceutical industry is reluctant to invest in the development of drugs to treat the major diseases of the poor because return on investment cannot be guaranteed. For example, for many years, eflornithine—a proven cure of human African trypanosomiasis (sleeping sickness)—was not produced due to market failure.[33] The disease surged, and patients had to be treated with worse, more toxic, and painful alternatives.[34]

Before any research is started, stakeholders should make sure implementation will not take away resources—including personnel, equipment, and health-care facilities—from other critical clinical and

---

[28] See Drugs for Neglected Diseases initiative, "Diseases and Projects."
[29] See Drugs for Neglected Diseases initiative, "Diseases and Projects."
[30] See Drugs for Neglected Diseases initiative, "Diseases and Projects."
[31] See World Health Organization, *Global Health Ethics*, 15.
[32] See World Health Organization, *Global Health Ethics*, 16.
[33] See Albert Sjoerdsma and Paul J. Schechter, "Eflornithine for African Sleeping Sickness," *Lancet* 354, no. 9174 (1999): 254.
[34] See Michael P. Barrett, "The Fall and Rise of Sleeping Sickness," *Lancet* 353, no. 9159 (1999): 1113–1134.

public health efforts.³⁵ This can be especially an issue in emergency situations. How will the rights and well-being of individual research participants be protected? In low-income countries, the only chance for medical care might be linked to participation in biomedical studies.³⁶ In these often-complicated circumstances, interests, and conflicts, well established integrity and distance are necessary. Experience has shown that researchers and research organizations cannot guarantee ethical trials by themselves, and unethical trials continue to be conducted.³⁷ Local research ethics committees should be established and independently assess the potential risk and benefits involved. Ongoing monitoring is necessary. In practice, global health research should be collaborative research. This means a fair sharing of data and samples between partners, the development of scientific capacity across the network, the allocation of scientific resources, joint decisions about authorship, and joint ownership of intellectual property.³⁸ According to the WHO,

> Individuals and communities that participate in research should, where relevant, have access to any benefits that result from their participation. Research sponsors and host countries should agree in advance on mechanisms to ensure that any interventions found to be safe and effective in research will be made available to the local population without undue delay, including, when feasible, on a compassionate use basis before regulatory approval is finalized.³⁹

ETHICAL DECISION-MAKING FRAMEWORK

Ethics involves judgements about the way we ought to live our lives, including our actions and intentions. The following ethical principles should be applied to any global health practice as a framework, helping to guide our actions. They are based on the World Health Organization's Global Health Ethics Unit recommendations.⁴⁰

- Justice or fairness (equity-fairness in the distribution of resources and outcomes, and procedural justice-fair process for making important decisions).
- Beneficence (acts that are done for the benefit of others, referring, in global health, to society's obligation to meet the

---

[35] See World Health Organization, *Guidance for Managing Ethical Issues*, 32.
[36] See World Health Organization, *Global Health Ethics*, 16.
[37] See Michael Carome, "Unethical Clinical Trials Still Being Conducted in Developing Countries," *The Huffington Post*, www.huffpost.com/entry/unethical-clinical-trials_b_5927660.
[38] See Michael Parker and Susan Bull, "Ethics in Collaborative Global Health Research Networks," *Clinical Ethics* 4, no. 4 (2009): 165–168.
[39] World Health Organization, *Guidance for Managing Ethical Issues*, 34.
[40] See World Health Organization, *Guidance for Managing Ethical Issues*, 8.

basic humanitarian needs such as nourishment, shelter, good health, and security).
- Utility (the rightness of actions are measured by the degree they promote the well-being of individuals or communities).
- Respect for persons (includes letting individuals make their own choices).
- Liberty (including religious and political freedoms).
- Reciprocity (consists of making a "fitting and proportional return" for contributions that people have made, and it is an important means of promoting the principle of justice).
- Solidarity (standing together; solidarity justifies collective action in the face of common threats).

In practice, setting up decision-making systems and procedures in advance is the best way to ensure that ethically appropriate decisions will be made. The more intrusive the proposed action, the greater the need for robust evidence that what is being proposed is likely to achieve its desired aim. When specific evidence is not available, decisions should be based on reasoned, substantive arguments and informed by evidence from similar situations, to the extent possible.[41]

Promoting global health ethics and principles, I want to voice some caution. I think of a friend of mine who spent a lot of time in the former Yugoslavia. He told me how local UN staff was complaining that under the old regime they had to learn Titoism and Marxism, while now it is human rights and ethics. For the majority of the local staff, who was looking for a job to feed their family and loved ones, it was the same indoctrination. While their boss and the message might have changed, the conditions remained the same.

We also have to accept that there will always be a downside to our action, even when, and because of, applying these principles. It may be just the dysfunction and imperfection of being human. However, not dealing openly with this limitation is unethical.

FINAL THOUGHTS

The humanitarian imperative requires responding to human suffering. We cannot accept that millions of people continue to die of curable and treatable diseases. At the same time, we should remember that it is ultimately the responsibility of governments to protect the health and well-being of their citizens. Humanitarian work is temporary and cannot be part of the permanent solution. Proximity to the patients and their suffering is fundamental. Keeping the individual at the center of the work also means that we should approach overwhelming or unimaginable problems without despair. Thirty-eight million people living with HIV/AIDS is an overwhelming global

---

[41] See World Health Organization, *Guidance for Managing Ethical Issues*, 9.

health problem; a twenty-two-year-old woman with AIDS is a human being who inspires compassion, who should be listened to and known. She can be treated, and her life be made better. We must strive to offer the best possible care, treatment, and prevention. This gives us the credibility to point out and address the root causes of the problem. Humanitarian action provides a human touch in an inhumane environment, and it may ultimately help to reestablish human dignity. Humanitarian global health is practiced according to universal ethics based on a moral approach that values human life and dignity. Such an approach is not utopian; it is very realistic ... and desperately needed. M

Nils Hennig, MD, PhD, MPH, is the Director of the Master of Public Health Program at the Mount Sinai School of Medicine and Associate Director of the Mount Sinai Global Health Center. Dr. Hennig—an expert in humanitarian aid, infectious diseases, and public health—has broad international health experience. He worked for the past fifteen years as physician, medical director, research coordinator, advisor, and medical consultant for *Médecins sans Frontières*, *Médecins du Monde*, MENTOR (Malaria Emergency Technical and Operational Response), the Fogarty International Center of the National Institute of Health, EarthRights International, Projecto Xingu, and other international organizations in humanitarian emergencies, fact finding missions, development, and research in the US and in many countries in the Global South—in Africa, Asia, Central America, the Caribbean, and South America. Dr. Hennig has a long record of training medical and public health staff of various international organizations and ministries of health in public health and infectious diseases. Dr. Hennig works clinically as attending at the Pediatric HIV/AIDS Clinic at Mount Sinai, in New York City, providing comprehensive care to infected/affected infants, children, adolescents, and young adults. He also continues international relief work, research, advocacy, and training for multiple agencies.

Conclusion

An Ethical Agenda for Global Public Health

Paul Farmer and Andrea Vicini, SJ

IN 1966, AT THE CONVENTION OF THE MEDICAL Committee for Human Rights in Chicago, Dr. Martin Luther King Jr. declared: "Of all the forms of inequality, injustice in health care is the most shocking and inhumane."[1] Tragically, King's words continue to ring true today. In New York City, one of the wealthiest cities in the world, people living in neighborhoods where more than 30 percent of residents are poor are 2.1 times as likely to die prematurely as those living in areas where less than 10 percent live in poverty. Brownsville in Brooklyn, East Tremont, Morrisania and Mott Haven in the Bronx, and Central Harlem have the highest premature death rates in New York City. Geographical location matters. If one compares the life expectancies of residents living in West 170th Street in the Bronx and those living less than seven miles away in West 96th Street in Manhattan, the gap is 10 years.[2]

How do we address such inequalities in health globally? First, we need to be aware that injustice is pervasive. Second, we must avoid desperation or complacent resignation; awareness of unacceptable disparities should animate social engagement to address and eliminate them. Third, it is necessary to make a preferential option for those who bear the brunt of these inequities. Scholars who embrace key tenets in Catholic social teaching, for example, do not shy away from calling such an engagement "a preferential option for the poor" and for the needy, the disadvantaged, and the marginalized.[3] In New York City, as elsewhere in the United States, these are overlapping social

---

[1] Martin Jr. Luther King, "Presentation at the Second National Convention of the Medical Committee for Human Rights, Chicago," (1966).
[2] See W. Li et al., *Summary of Vital Statistics 2017: The City of New York* (New York: New York City Department of Health and Mental Hygiene, Bureau of Vital Statistics, 2019).
[3] See Stephen J. Pope, "Proper and Improper Partiality and the Preferential Option for the Poor," *Theological Studies* 54, no. 2 (1993): 242–271; Stephen J. Pope, "Christian Love for the Poor: Almsgiving and the 'Preferential Option'," *Horizons* 21, no. 2 (1994): 288–312.

categories including a disproportionate fraction of Black and Brown people.

In such a way, theological ethical discourse (and the praxis that it promotes) provides us with the guiding principles for individual and social engagement in society, including promoting health across the planet. For practitioners within medicine and public health, the preferential option is a moral and methodological guide to achieving health equity.[4] Writing from Marquette University and sharing the voices and stories of people in the Global South facing great distress and poverty, the Brazilian theologian Alexandre Martins reminds us that, "The preferential option for the poor is at the heart of liberation theology as well at the heart of our approach to social justice in health care."[5]

PREFERENTIAL OPTION FOR THE RICH AND WHITE

Examining what it means to make a preferential option for the poor in global public health invites us to revisit how a preferential option has more commonly been made for the rich and white. An option for the privileged has burdened our past and continues to influence our present. Colonial history provides us with an example.

Sir Ronald Ross was a British medical doctor, born in India when it was the world's largest colony, who joined the Indian Medical Service in 1881. Among other interests, he focused on malaria and its then still unknown pathogenesis. Based on the studies of the French physician Charles Louis Alphonse Laveran[6] and the Scottish physician Patrick Manson, Ross hypothesized that mosquitoes were linked to the propagation of the disease.[7] After a few years of failed studies, Ross succeeded in demonstrating the life-cycle of malaria parasites in mosquitoes. For his work elucidating the transmission of malaria, he received the Nobel Prize for Physiology or Medicine in 1902.

To study malaria, Ross traveled to one of the most malaria-ridden British colonies in West Africa: Freetown, the capital of Sierra Leone. There, he helped implement what he considered to be sound public health interventions. Ross suggested means of lessening transmission by attacking the vectors with larvicide and screens. He also suggested

---

[4] See Michael P. Griffin and Jennie Weiss Block, eds., *In the Company of the Poor: Conversations between Dr. Paul Farmer and Fr. Gustavo Gutierrez* (Maryknoll, NY: Orbis Books, 2013).
[5] Alexandre A. Martins, *The Cry of the Poor: Liberation Ethics and Justice in Health Care* (Lanham, MD: Lexington Books, 2019), 60.
[6] In 1907, Alphonse Laveran won the Nobel Prize in Physiology or Medicine for his research on how protozoa cause diseases.
[7] Patrick Manson discovered that insects can be host to developing parasites, which can cause a disease in human beings. In particular, he revealed that mosquitos can host the worm *Filaria bancrofti* and lead to filariasis, when the worm invades body tissues. He is considered the founder of the field of tropical medicine.

segregating the city, by establishing residential areas for the rich, white colonizers in the Hill Station, on the hills overlooking the city, while locals were left at the bottom of the hills and given neither the means to protect themselves from malaria nor permission to travel to the hilly residential area. (Evidence that mosquitoes could fly unbound, disregarding any racial and residential segregation, did not deter Dr. Ross.) What we might call an option for the rich and white, in other words, provided the framework for Ross's policy prescriptions.

PREFERENTIAL OPTION FOR THE EMPIRE

Colonial medicine's option for the rich and white continues to structure moral failures in global public health. Critical attention to uneven power dynamics, and how they are historically determined, is thus sorely needed, and clarifies what might be termed a preferential option for the empire.

As humankind now faces another global pandemic, this one caused by a novel coronavirus, it seems relevant to reflect on the last "big one"—the 1918 influenza pandemic. The most widespread pandemic of the 20th century was caused by an H1N1 virus with a few genes of avian origin. Historical accounts estimate that about 500 million people were infected, one third of the world's population, with about 50 million deaths worldwide. No vaccine or antibiotics to treat secondary bacterial infections were available. The spread of infection could be controlled only through isolation, quarantine, personal hygiene, disinfectants, limiting public gatherings, and delivering supportive and critical care.

On August 1, 1918, the HMS Mantua, a passenger ship of the British Royal Navy converted into armed merchant cruiser, sailed from Plymouth, Great Britain, to Freetown, Sierra Leone, arriving on August 15. A growing number of its sailors were ill with influenza.[8] Within a month of the Mantua's arrival, 4 percent of Freetown citizens had died. Moreover, the viral infection spread rapidly in the country and across the continent via railway and shipping lines.[9] Hence, the Mantua is considered one of the first ships to have propagated the spread of the 1918 pandemic to the African continent.[10]

---

[8] See Paul Farmer, "Ebola, the Spanish Flu, and the Memory of Disease," *Critical Inquiry* 46, no. 1 (2019): 56–70, at 58.
[9] See ibid. 61 and 62. See also K. David Patterson and Gerald F. Pyle, "The Diffusion of Influenza in Sub-Saharan Africa during the 1918-1919 Pandemic," *Social Science & Medicine* 17, no. 17 (1983): 1299–1307, at 1302–1303.
[10] See John M. Barry, *The Great Influenza: The Epic Story of the Deadliest Plague in History* (New York: Penguin Books, 2005), 182; Connie Goldsmith, *Influenza* (Minneapolis, MN: Twenty-First Century Books, 2011).

Colonial health authorities relied on a readily recognized playbook, and it was punitive.[11] Just a few months after its arrival in Freetown, the Nigerian newspaper *Lagos Standard*, on October 2, 1918, reported, "People are hustled out to practically certain death in a building where... those sent there are obliged to lie on bare cement floor."[12] Moreover,

> It is not a wise thing to depend on Force as the most essential weapon for stamping out an epidemic. The cooperation of the people ... is very essential and that cooperation cannot be secured by the present methods... which make the people run away not from dread of the disease, but from fear of sanitary officials and their ways.[13]

While one might rightly consider such methods worse than punitive, since there was little care given in "buildings" mislabeled as "treatment centers" or "hospitals," some of the same practices were applied during the 2013–2016 Ebola pandemic in Guinea, Liberia, and Sierra Leone. It is a recurring pattern that could be defined as a "preferential option for the artists formerly known as the empire." It is the empire of colonial medicine, with its classification of worthy and unworthy citizens, that furthers discrimination and dehumanization in health care.

In ordinary times, and even more so in emergencies, the cooperation and trust between citizens and healthcare professionals is essential, but it is not a given. Successful and trusting interactions depend, first, on the critical assessment and denunciation of any colonialist logic that does not consider citizens as equal participants, consequently hampering any public health measures; and, second, on building and protecting the trust and social solidarity needed to curtail fear of public health officials and of the sanitary or social measures they propose.

Imperial logic has a great cost among the most vulnerable. In 2014, in West Africa, 70 percent of people who got sick with Ebola died because the healthcare systems were far too weak to respond to a fast moving epidemic. At the same time, no American citizens afflicted with Ebola and treated in US hospitals died. The disparity highlights the unequal distribution of quality medical care and the ways in which social forces, including poverty and racism, dictate who lives and who

---

[11] See Paul Farmer, *Fevers, Feuds, and Diamonds: Ebola and the Ravages of History* (New York: Farrar, Straus and Giroux, 2020).

[12] "Isolation Camp Abekun," *Lagos Standard*, October 2, 1918, 6.

[13] "Influenza in Lagos," *Lagos Standard*, October 2, 1918, 4. Quoted in Farmer, "Ebola, the Spanish Flu, and the Memory of Disease," 64. See also Sandra M. Tomkins, "Colonial Administration in British Africa During the Influenza Epidemic of 1918-19," *Canadian Journal of African Studies—Revue Canadienne des Etudes Africaines* 28, no. 1 (1994): 60–83.

dies. In West Africa, in 2014 as in 1918, sick and vulnerable patients were isolated without care. We need to critically examine these inequities and resolutely engage ourselves in the fight to strengthen healthcare systems, striving for more inclusive access to health services for those who lack them across the globe.

WHERE THERE IS NO NURSE (OR DOCTOR): EBOLA AS A CAREGIVERS DISEASE

When there are not enough doctors, nurses, or other healthcare professionals to care for the sick, the family and traditional healers (many of them kin and most of them socially proximate) are often the sole caregivers. Families greatly suffer as a result. Ebola and other infectious diseases can disproportionately affect close contacts. During and after colonial rule, but mostly after, medical and nursing schools to train healthcare professionals were not established. Moreover, health workers in the Global South are often "pushed" out of their jobs by the absence of the tools of their trade and "pulled" into settings in the Global North instead.[14]

SOCIALIZATION FOR SCARCITY ... ON BEHALF OF OTHERS

Public health crises exacerbate inequities already present in the social fabric, including racial discrimination, economic marginalization, and inequalities in housing and education. As major social institutions, universities bear a great responsibility to foster individual and collective flourishing and promote the common good.[15] Both in research and teaching, universities should reinforce their commitment to engage in reparative work that attempts to correct past inequities and tend toward greater equity. On the one hand, university scholars can continue to help health practitioners learn from the history of global public health with its harms and achievements. On the other hand, a strong commitment to social flourishing requires collaborative engagements and pragmatic commitments to strengthen healthcare systems, to empower those made vulnerable, and to enact

---

[14] As an example, see Ruth Groenhout, "The 'Brain Drain' Problem: Migrating Medical Professionals and Global Health Care," *International Journal of Feminist Approaches to Bioethics* 5, no. 1 (2012): 1–24; Alina Botezat and Raul Ramos, "Physicians' Brain Drain: A Gravity Model of Migration Flows," *Global Health* 16, no. 1 (2020): 7; Yusuf Yuksekdag, "Health without Care? Vulnerability, Medical Brain Drain, and Health Worker Responsibilities in Underserved Contexts," *Health Care Analysis* 26, no. 1 (2018): 17–32.

[15] See Kristin E. Heyer, "The Idea of the Common Good: Interdisciplinary Contributions to Catholic Higher Education," *Integritas* 7, no. 1 (2016): 1–20; Kristin E. Heyer, "Practicing Intellectual Hospitality: The Common Good and the Work of the Jesuit University," *Explore* 20 (2018): 8–18; David Hollenbach, "The Catholic University under the Sign of the Cross: Christian Humanism in a Broken World," in *Finding God in All Things: Essays in Honor of Michael J. Buckley, SJ*, ed. Michael J. Himes and Stephen J. Pope (New York: Herder and Herder, 1996), 279–298.

purposively reparative actions. One essential goal must be to support institutions that train health professionals in the Global South, who in turn care for their local communities. An example is the University of Global Health Equity in Rwanda,[16] a collaborative initiative of the medical nonprofit Partners In Health and the Government of Rwanda. Many similar engagements are needed.[17]

In conclusion, an ethical agenda for global public health must be shaped and informed by the preferential option for the poor. Its implementation demands that we revisit our histories and requires our collaborative engagement, accompanying the poor and the otherwise marginalized to bring justice into the world. **M**

Paul E. Farmer, MD, PhD, is Kolokotrones University Professor of Global Health and Social Medicine in the Blavatnik Institute, Department of Global Health and Social Medicine, at the Harvard Medical School. With his colleagues in the US and abroad, he pioneered novel, community-based treatment strategies that demonstrate the delivery of high-quality health care in resource-poor settings in the US and other countries. This work is documented in prestigious journals. An engaging speaker, Dr. Farmer has written extensively on health and human rights, the role of social inequalities in the distribution and outcome of infectious diseases, and global health. His most recent books are *Fevers, Feuds, and Diamonds: Ebola and the Ravages of History* and *In the Company of the Poor: Conversations with Dr. Paul Farmer and Fr. Gustavo Gutiérrez*. Other titles include *To Repair the World: Paul Farmer Speaks to the Next Generation*; *Haiti After the Earthquake*; *Partner to the Poor: A Paul Farmer Reader*; *Pathologies of Power: Health, Human Rights, and the New War on the Poor*; *Infections and Inequalities: The Modern Plagues*; *The Uses of Haiti*; and *AIDS and Accusation: Haiti and the Geography of Blame*. In addition, Dr. Farmer is co-editor of *Reimagining Global Health: An Introduction*; *Women, Poverty, and AIDS: Sex, Drugs and Structural Violence*; *The Global Impact of Drug-Resistant Tuberculosis*; and *Global Health in Times of Violence*.

Andrea Vicini, SJ, is Michael P. Walsh Professor of Bioethics and professor of Moral Theology in the Theology Department at Boston College. He is an alumnus of Boston College (STL and PhD) and holds a MD from the University of Bologna and an STD from the Pontifical Faculty of Theology of Southern Italy in Naples. At Boston College, he was Gasson Professor and taught as the School of Theology and Ministry. He also taught in Italy, Albania, Mexico, Chad, and France. He is co-chair of the international network Catholic Theological Ethics in the World Church, as well as lecturer and member of associations of moral theologians and bioethicists in Italy, Europe, and the US. His research interests and publications include theological bioethics, global public health, new biotechnologies, environmental issues, and fundamental theological ethics.

---

[16] See "University of Global Health Equity," ughe.org/.
[17] See "Partners in Health," www.pih.org/.

www.ingramcontent.com/pod-product-compliance
Lightning Source LLC
Chambersburg PA
CBHW070327230426
43663CB00011B/2247